# Advanced LaTeX in Academia

Marco Öchsner · Andreas Öchsner

# Advanced LaTeX
# in Academia

## Applications in Research and Education

 Springer

Marco Öchsner
University of Cambridge
Cambridge, UK

Andreas Öchsner
Faculty of Mechanical and Systems
Engineering
Esslingen University of Applied Sciences
Esslingen, Germany

ISBN 978-3-030-88958-6          ISBN 978-3-030-88956-2   (eBook)
https://doi.org/10.1007/978-3-030-88956-2

This Springer imprint is published by the registered company Springer Nature Switzerland AG
The registered company address is: Gewerbestrasse 11, 6330 Cham, Switzerland

*"I can't go to a restaurant and order food because I keep looking at the fonts on the menu. Five minutes later I realize that it's also talking about food. If I had never thought about computer typesetting, I might have had a happier life in some ways."*

*Donald Knuth, creator of TeX*

# Preface

This book contains a comprehensive treatment of advanced LaTeX features. The focus is on the development of high quality documents and presentations, by revealing powerful insights into the LaTeX language. The well-established advantages of the typesetting system LaTeX are the preparation and publication of platform-independent high-quality documents, with automatic numbering and cross-referencing of illustrations or references. The functionality of these documents can be extended in various ways, particularly in conjunction with the capabilities of the portable document format (PDF) and various programming languages, generating highly dynamic and flexible electronic documents.

Cambridge, UK                                                    Marco Öchsner
Esslingen, Germany                                            Andreas Öchsner
August 2021

**Acknowledgements** We would like to express our sincere appreciation to Springer Nature, especially to Dr. Christoph Baumann (Editorial Director), for giving us the opportunity to realize this book.

# Contents

# Chapter 1
# Introduction: The Basics

**Abstract** This chapter will bring the reader up to speed with the basic usage of LATEX. This will be a fast-paced introduction to the fundamental structure of LATEX documents, and the most frequently used commands.

## 1.1 Where to Find Help

Before getting started with this chapter, we present some key resources here where to find help when first starting out with LATEX, see Table 1.1.

## 1.2 Anatomy of a LaTeX Document

LATEX is a typesetting system that works as a high-level markup language. This means that instead of having to 'click' your way through formatting options and the 'what you see is what you get' approach many document processing programs implement, one uses a rich set of commands to specify a document's components (titles, sections, bold font, italics, figures, among others), and lets LATEX take care of the formatting. Of course, you can additionally personalise and further specify formatting and features of the document, which is what this book is largely about. To begin with, we will briefly review the basic structure and function of a LATEX document. If you have used LATEX this should all be quite familiar to you, and you might find it more useful to skip to the next chapter.

A LATEX document (a *.tex document file) is split into two main parts: the *preamble* and the main body of the document. The preamble contains all *global* specifications and formatting settings, such as paper size, font size, what kind of document it is (article, book, report, letter...). It is also where we include 'packages', which allow access to libraries of macros to add functionality to LATEX. Then comes the main body of the document, often divided into different parts, chapters, or sections.

M. Öchsner and A. Öchsner, *Advanced LaTeX in Academia*,
https://doi.org/10.1007/978-3-030-88956-2_1

**Table 1.1** Some web resources to start with LaTeX

| Site | URL |
|------|-----|
| LaTeX Wikibooks | https://en.wikibooks.org/wiki/LaTeX |
| Overleaf tutorials | https://www.overleaf.com/learn/latex/Tutorials |
| TeX stackexchange | https://tex.stackexchange.com/ |
| LaTeX subreddit | https://www.reddit.com/r/LaTeX/ |

The preamble begins with the command

```
\documentclass[]{}
```

The curly brackets allow you to specify the document type (or 'class') - the standard ones being `article, book, report` and `letter`. After this, the main document begins and ends with

```
\begin{document} ... \end{document}
```

A minimal example of a LaTeX document thus is:

```
1  \documentclass{article}
2    \begin{document}
3        Hello, World!
4    \end{document}
```

**Listing 1.1**  Minimal example of a LaTeX document

The simplest way to get a typeset document in PDF from this, would be to save this file, e.g. as `example.tex` and to then compile it from terminal by running `pdflatex example.tex` (this requires installation of pdflatex, which we will cover later in the chapter). This would then generate a file called `example.pdf` in the same folder.

### 1.2.1  Document Classes

There are some key differences between the document classes, apart from their layout, with regards to default settings and available commands:

- The standard sectioning commands in LaTeX are `\part{}`, `\section{}`, `\subsection{}`, and `\paragraph{}`. These are all common to the three standard document classes, while `report` and `book` additionally have a higher sectioning command `\chapter{}`, and `article` has an additional lower sectioning command `\subsubsection{}`.
- The document classes `book` and `report` begin with a titlepage (if specified) as a standalone page, while article does not.

- The numbering of figures and other floats, as well as equations, occurs for each chapter in `book` and `report`, while it runs globally for `article`.
- The `book` document class specifies `\frontmatter`, `\mainmattter` and `\backmatter` commands, which change page numbering for the main body (regular arabic numerals), and the pages outside this (roman numerals before the main body, none after).
- The `\abstract{}` environment is not available in the `book` document class.

The square brackets in the `\documentclass[]{}` allow us to specify (optional) additional settings. These are:

- paper size. Standard options include: `a4paper`, `a5paper`, `b5paper`, `letterpaper`, `executivepaper` and `legalpaper`. The default page size is US letter, similar to A4. It is often best to specify special page sizes with the `geometry` package, which will be covered later in the chapter.
- standard font size, e.g. `12pt`. For the standard document classes, `10pt` (default), `11pt`, and `12pt` are accepted. Further font size specification can be achieved in the document body (or with other document classes).
- whether the document is single- or double-sided, with `oneside` or `twoside`. By default, `book` is `twoside`, and `article` and `report` are `oneside`.
- For `book` and `report`, the default option `openright` forces chapters to always begin on a right-hand, odd page, and `openany` permits chapters and parts to open on the left hand page.
- draft mode, where no pictures are compiled (allows quicker compilation) - with `draft`.

It is important to remember that these just set the standard global settings - in the main body of the document, you can locally change these when required.

The beginning of a LaTeX document might thus be, for example:

- `\documentclass{article}` The simplest start, selecting all defaults for the article document class.
- `\documentclass[a4paper,oneside,11pt]{book}` For a document typeset as a book with A4 page dimensions, 11 pt default text font size, and that is one-sided (i.e. consistent margins and header and footers for all pages).
- `\documentclass[10pt]{report}` For a document typeset as a report, with a standard text size of 10 pt.

After specifying the document class and desired options, the preamble also allows us to call packages, so that we can access macros for specific function in LaTeX. The command to include these is

$$\texttt{\textbackslash usepackage[]\{\}}$$

Popular packages include `geometry`, `graphicx`, `amsmath`, `xcolor`, among many others. A database of packages, with their documentation, can be found at www.ctan.org. Table 1.2 gives an overview of the most popular packages and their uses.

**Table 1.2** Popular LaTeX packages

| Package | Usage |
|---------|-------|
| graphicx | Necessary to include images in the document. Provides the includegraphics command |
| babel | Changes default language of the document (useful for correct hyphenation and linebreaks) |
| amsmath, amssymb | Supplies a comprehensive set of commands to format mathematical statements, as well as symbols |
| hyperref | Hyperlink management in LaTeX. Also allows to link to references in the text to figures, tables, and other items. Covered in Chap. 2 |
| geometry | Allows to set up page definitions as desired, controlling features such as margins and text widths |
| microtype | Control over kerning and other font settings |
| booktabs, multicol, tabularx, tabulary | Advanced table formatting |
| siunitx | Automated rendering of numbers in SI units and significant figure handling |
| tikz, pgfplots | Plotting and drawing packages. Extensively used in Chap. 3 |
| fancyhdr | Constructing headers and other advanced page formatting |
| xcolor | Allows definition and selection of colors |
| biblatex, natbib | Bibliographical management in LaTeX. See more in Chap. 2 |

A first modification often required is to adjust the page margins and line spacing. To do this, the `geometry` package is very useful. Not only can margins be specified in a variety of units, but also many other parameters. Margins can either be set with `\usepackage[margin=1.5cm]{geometry}`, or by specifying each margin individually, with

```
\usepackage[left=2cm, right=2cm, top=1.54cm, bottom=1cm]{geometry}
```

A simple way of changing line spacing is by using the `setspace` package. By having `\usepackage[...]{setspace}` in the preamble, we can for example set `[doublespacing]` to obtain a double spaced document, or

```
\usepackage[onehalfspacing]{setspace}
```

for 1.5x line spacing. Finally, in the preamble we can give some information for the document class to generate a title page. The command `\maketitle` generates a basic title page in most document classes, except in the `article` document class, where a title is generated at the top of the page. A simple example is:

```
1  \documentclass{article}
2      \title{An Introduction to \LaTeX}
3      \author{Authors}
4      \date{2021}
5  \begin{document}
6  \maketitle
7
8  \end{document}
```

**Listing 1.2**  Example of generating a title page

## 1.2.2  Basic Environments

### 1.2.2.1  Figures

Figures can be included in a LaTeX document simply by using the

$$\verb|\includegraphics[]{}|$$

command, if we include the graphicx package in the preamble with `\usepackage{graphicx}`. In the square brackets, sizing options can be passed—for example, with `[scale=1.3]` we can change the figure's size by a specified factor. We can also scale a figure to have a specified width, or relative to the text's width, such as with

$$\verb|\includegraphics[width=0.8\textwidth]{figure.eps}|$$

which will scale the image to 80% of the text width. We can also fix the height of a figure, or both width and height, e.g. with `[width = 2cm, height = 5 cm]`. Images can also be easily rotated. For example, with `[angle=90]` we can simply rotate a figure by 90 degrees.

In curly brackets, we have to pass the filename of our figure. If the figure is in the same folder as our main document, then the filename itself will suffice; else, we must include the folder path to the image. File extensions can be omitted. A default location for LaTeX to search for figures (outside the directory of the main document), can be set globally in the the preamble, for example, with

$$\verb|\graphicspath{{./figures}{./additionalFigures}}|$$

The '.' here refers to the current working directory. While in Unix systems, we specify folders using absolute or relative paths, in Windows we can set paths to folders with

$$\verb|\graphicspath{{C:/Administrator/Desktop/Document_Figures/}}|$$

By placing the \includegraphics[]{} command within the \begin{figure}... \end{figure} environment, we can specify its positioning relative to the page, as well as give the figure a caption and label. We can tell LaTeX to put the figure more or less at the same location in the text as it is located in the source code, by using \begin{figure}[h]. If we want the figure to be placed on top, or bottom, of the page, we can substitute the h by t, or b; using p allows us to have it on its own page, and adding an ! overrides LaTeX default positioning parameters. The default alignment of a figure in LaTeX is on the left of the page. As often images are to be centred, placing the command \centering in the figure environment achieves this.

The positioning of the caption, above or below the figure, depends simply on whether the \caption{...} command is above or below the \includegraphics[]{} command. Labels are useful, as we can use them to refer to a figure (or table, equation, among others) in the text, and LaTeX will automatically substitute it by the correct figure number.

```
1  \begin{figure}
2      \centering
3      \includegraphics{figure1}
4      \caption{This is a caption}
5      \label{fig:figurespslabel1}
6  \end{figure}
```

**Listing 1.3** Example of a figure environment with a caption

In some cases, it might be desirable to have text wrapping around a figure. For this purpose, the package wrapfig can be used - by including

$$\usepackage\{wrapfig\}$$

in the preamble of our document. Here, we do not use \begin{figure}, but insert wrapped figures with \begin{wrapfigure}{alignment}{width}...\end{wrapfigure}. For alignment, we can set r or l for right or left aligned figures, and for width any of the commands used in the regular figure environment can be used. The actual figure itself is inserted as before, such that we can use:

```
1  \begin{wrapfigure}{l}{0.3\textwidth}
2      \centering
3      \includegraphics[width=0.3\textwidth]{figure}
4  \end{wrapfigure}
```

**Listing 1.4** Minimal example of wrapfigure environment

In this case, our figure will be aligned on the left, and be 30% of the text width wide. Note that the width of the figure is not relative to the wrapfigure box, but still relative to the text width, and thus we have set their widths to be identical, such that the figure fills up the wrapfig space.

**Table 1.3** Sample table

| cell 11 | cell 12 | cell 13 |
|---------|---------|---------|
| cell 21 | cell 22 | cell 23 |
| cell 31 | cell 32 | cell 33 |

#### 1.2.2.2 Tables and Tabulars

LATEX provides a number of packages and tools to generate tables. Here, two environments are used together, the `table` and `tabular` environments, where the first provides the positioning, alignment, label and caption of the table, and the latter the table itself. It should be noted that the `table` environment is not required, and one can simply insert a table with `\begin{tabular}` and `\end{tabular}`, but this allows very few adjustments. A simple example is shown in Listing 1.5:

```
1  \begin{table}[!h]
2  \centering
3     \begin{tabular}{ lll }
4     cell 11 &  cell 12 & cell 13  \\
5     cell 21 & cell 22  &  cell 23\\
6     \end{tabular}
7  \caption{Table caption}
8  \label{tab:myspstable}
9  \end{table}
```

**Listing 1.5** Minimal example of a LATEX table

As in the `figure` environment, the positioning parameter in square brackets after `\begin{table}` allows the table to be placed either at the top, on its own, at the bottom, or as is most common, at that exact position in the text as it is located in the source code. `\centering` aligns the table to the center of the text. As before, the caption command, if placed before or after the `tabular` environment, will print a table caption above or below the table. In the `tabular` environment, the parameters in the curly brackets next to `\begin{tabular}` indicate the alignment of the table cells (l - left, r- right, c - center). A particular column width can also be defined using p{<width>}. Including vertical bars will draw table borders between the indicated columns, such that { |c|c|c| } will generate all vertical column borders. Using { ||c|| } will result in two border lines being drawn. To draw horizontal separating lines, the command `\hline` must be placed after each row (Table 1.3).

```
1   \begin{table}[!ht]
2   \centering
3      \begin{tabular}{p{2cm}|p{3cm}p{3cm}}
4      cell 11 & cell 12 & cell 13 \\ \hline
5      cell 21 & cell 22 & cell 23 \\
6      cell 31 & cell 32 & cell 33
7      \end{tabular}
8   \caption{Sample table}
9   \label{tab:myspstable1}
10  \end{table}
```

**Listing 1.6** LATEX table with specified column width

For more advanced table formatting options, the package `booktabs` provides further customisation options. To rotate tables, the package `lscape` provides the environment `\begin{landscape} ...\end{landscape}`. For tables going across multiple pages, the package `longtable` can be used.

### 1.2.2.3   Lists

LaTeX has a variety of list-type environments. The most basic of these are the `enumerate` and `itemize` environments, which allow lists that are numbered, or utilise bullet points (or other symbols) to be created. Each list item begins with an `\item` command.

```
1  \begin{itemize}
2      \item Sample item 1
3      \item Sample item 2.
4  \end{itemize}
```

**Listing 1.7** Basic example of a LaTeX list

- Sample item 1
- Sample item 2.

We can combine both these environments to generate nested lists, such as:

```
1  \begin{itemize}
2      \item Sample item 1
3      \begin{enumerate}
4          \item Numbered item 1
5          \item Numbered item 2
6      \end{enumerate}
7      \item Sample item 2
8  \end{itemize}
```

**Listing 1.8** Basic example of a LaTeX nested list

- Sample item 1
  1. Numbered item 1
  2. Numbered item 2
- Sample item 2

Numerated lists can also be nested, though here the counter will change, going from arabic numerals to lowercase letters, to Roman numerals.

Many more options exist to format lists. In the itemize environment, the bullet can be changed by placing the desired symbol in square brackets just after `\item`; for example, a dash can be used with `\item[--]`, while an asterisk can be used with `\item[$\ast$]`.

Other packages allow for extended functionality, such as the package `etaremune` generates lists with reverse numbering.

#### 1.2.2.4  Mathematical Expressions

There are a number of math-related environments in LaTeX. A simple way to add in-line math commands, is to simply use the dollar sign to indicate the start and end of a mathematical expression, as in

$$\$\frac{2\pi}{12}\$$$

Alternatively, one can use  `\( ... \)` or `\begin{math} ...\end{math}`. Equations that are to be displayed on their own, can be included by using two ampersands, as in

$$\&\&y=\frac{3x}{2\epsilon}\&\&$$

or by using a backslash with square brackets, as in `\[ ... \]`. The `equation` environment allows for longer mathematical expressions to be generated. The following commands will generally require use of one of the aforementioned math environments.

Some of the more commonly used math commands used are those for fractions, that are generated using the `\frac{}{}` command, for example

$$\frac{3x}{x}$$

Exponents and indices can be generated using ^ and _ respectively, with the caveat that if one wants to super- or subscript more than one character, one must put these in curly brackets, such as

$$x^{m+n} = x^m \cdot x^n$$

which prints $x^{m+n} = x^m \cdot x^n$.

Greek characters are generally accessible in a math environment, using for example `\alpha` or `\Alpha`, with the command starting with capital letter generating a capital Greek letter. In general, the lower case letters will be italicised automatically. For upright Greek letters, the package `\usepackage{upgreek}` can be used, with the corresponding Greek letters generated with the commands `\upalpha` or `\Upalpha`. For the capital sigma used in summation, we can use the commands `\sum`, using the same syntax for super and subscripts to indicate lower and upper limits, e.g. `\sum_{n=1}^\10`. A similar method is used for the capital pi used for products using `\prod`, and for integrals, using `\int`.

Square roots can be generated using `\sqrt[]{}`, with the nth root indicated in square brackets. Vector arrow notation can be used with either `\vec{a}` or `\overrightarrow{b}`, producing $\vec{a}$ and $\overrightarrow{b}$ respectively. For most mathematical symbols, there is a corresponding LaTeX command. Some examples: the infinity symbol - `\infty` - $\infty$; the greater and lesser equal to signs - `\geq` and `\leq`

respectively; dot notation for derivatives can be used with \dot{} and \ddot{};
the commands \forall, \exists, \in, \mapsto produces the symbols they
clearly refer to. Mathematical fonts also can be used in LaTeX with \mathbb{R}
producing $\mathbb{R}$, and \mathcal{T} for $\mathcal{T}$. Some examples of these in use:

```
1  $(f \circ f^{-1})(n) = n$
2  $\sum\limits_{i=3}^{10} i^2$
3  $x = -\frac{b}{2a} + \sqrt{\frac{b^2-4ac}{4a^2}}$$
4  $x_{1,2} = \frac{-b \pm \sqrt{b^2-4ac}}{2a}$
5  $\overrightarrow{AD} = \frac{1}{3}(-5a+2b)$
6  $\forall n \in \mathbb{N}   : (\exists m \in \mathbb{N}   : m \geq n)$
```

**Listing 1.9**  Examples of math expressions

Which results in the following expressions:

1. $(f \circ f^{-1})(n) = n$
2. $\sum\limits_{i=3}^{10} i^2$
3. $x = -\frac{b}{2a} + \sqrt{\frac{b^2-4ac}{4a^2}}$
4. $x_{1,2} = \frac{-b \pm \sqrt{b^2-4ac}}{2a}$
5. $\overrightarrow{AD} = \frac{1}{3}(-5a + 2b)$
6. $\forall n \in \mathbb{N} : (\exists m \in \mathbb{N} : m \geq n)$

## 1.3  Getting Started

To get started with compiling LaTeX documents, one must have at least a LaTeX
compiler installed. Usually, users will also have a preferred LaTeX editor installed,
acting as interface where one can type and edit documents, and initiate compilation.
Currently, the most commonly used LaTeX compiler is pdflatex. This is included in
the MikTex package https://miktex.org/, or in the TeXlive package http://www.tug.
org/texlive/. When installing MikTex, a basic LaTeX editor is provided, TeXworks.
Nevertheless, the simplest way might be to use an online editor when starting out,
such as www.overleaf.com or www.cocalc.com. An overview of different LaTeX
compilers and editors will be given in the next chapter.

# Chapter 2
# Advanced Formatting

**Abstract** This chapter addresses advanced formatting options, such as reference management, page setup, and font management. Furthermore, the creation of new commands and integration with other formats and programming languages is introduced. The chapter closes with some comments on cross-referencing, hyperlinks and additional tools.

## 2.1 Reference Management

Citing and generating bibliographies may appear daunting at first in LaTeX. While there are many plugins and reference managers that easily integrate with popular word processing programmes, references in LaTeX may not appear as straightforward. In this section, we will cover the various ways one can manage references and cite in LaTeX documents.

For documents not requiring many references, it may suffice to provide the bibliography within the LaTeX document itself. An example of this is given in Listing 2.1, showing references listed in a `thebibliography` environment, which creates a 'Bibliography' section in the `book` and `report` document classes, and a 'References' section in the `article` document class. Each entry is added with the command `\bibitem`, with an identifying key (that will be used when citing) given in curly brackets, and a label given in the square brackets, that will be displayed in the citation as well as in the bibliography. Without a label, items cited will simply be numbered in square brackets. Each item in the bibliography can be cited using the command `\cite{key}`. The parameter 99 in the `thebibliography` environment indicates the 'widest' number to be used in bibliography numbering. The result of this example bibliography, when citing both references, is shown in Fig. 2.1, when using the `article` document class.

```
1  \begin{thebibliography}{99}
2    \bibitem[displayLabel1]{key1} J. Smith. \emph{Book Title}, Publisher,
3      City, 1970.
4    \bibitem[displayLabel2]{key2} J. Doe. Article Title. \emph{Journal Title},
5      100(75):25–50, 1980.
6  \end{thebibliography}
```

**Listing 2.1** Minimal example of bibliography in LaTeX

# References

[displayLabel1]  J. Smith. *Book Title*, Publisher, City, 1970.

[displayLabel2]  J. Doe. Article Title. *Journal Title*, 100(75):25-50, 1980.

**Fig. 2.1**  Example of a basic 'References' section, see Listing 2.1

While this approach may prove sufficient for smaller projects, for any larger works and documents requiring large quantities of references, it is advisable to first generate a .bib file containing all the required references. This file must be formatted according to a particular pattern as shown in Listing 2.2, and can then be used to cite in LATEX. LATEX can then access this file to generate the adequate citations and references, using either BibTeX or biber. The most common packages used to organise this process (i.e. containing the macros we will use) are the biblatex or natbib packages.

To generate this .bib file, there are two main options: using a reference manager that allows exporting a .bib file (thanks to the wider adoption of LATEX, popular reference managers such as Zotero, Mendeley, or EndNote have such functions). Alternatively, one can create a .bib file manually. This may be of some use when downloading citations in the .bib format, as many databases offer this (e.g. NCBI PubMed). All this requires is a simple text file containing a list of the desired references in the .bib format, as seen in the Listing 2.2. The type of entry is indicated with an @ (e.g. @book), and the entry then given within curly brackets. The first field that must be provided is the key (used when citing), and fields are separated by commas. Note that author names are given with <Last Name, First Name>, and separated with an 'and' (e.g. with multiple initials, Brown, T. A. and Hamilton, T. F. - note the space between initials); when an institution or organisation is given in the author field, it is best to enclose it in an additional set of curly brackets, to avoid automatic shortening of the first 'name', and for BibTeX to obey the capitalisation indicated.

```
1   @book{Smith2021,
2        author = {Smith, John},
3        title = {Formatting Bibliographies in LaTeX},
4        publisher = {Springer},
5        city = {Heidelberg},
6        year = {2021}}
7
8   @article{JonesSmith2013,
9        author = {Jones, Angela and Smith, William},
10       title = {A Groundbreaking Journal Paper},
11       journal = {Journal Name},
12       volume = {35},
13       issue = {4},
14       pages = {122–135},
15       year = {2013}}
```

**Listing 2.2**  Typical .bib file entry

The fields that are available for a `bib` entry, depend on the kind of entry one specifies. All `article` entries require the fields author, title, year, and journal; any `book` entry, similarly, requires author or editor, title, publisher, and year. Other fields are available for these, such as volume, number, pages, among other fields. To cite web pages, very often the entry type `misc` is used, as in the example below.

```
1  @misc{WordLatexNature2019,
2      title        = "Craft beautiful equations in Word with LaTeX",
3      author       = "{Matthews, David}",
4      howpublished = "\url{https://www.nature.com/articles/d41586-019-01796-1}",
5      year         = 2019,
6      note         = "Accessed: 2021-06-01"}
```

**Listing 2.3** Miscellaneous .bib entry

Once a file with all the necessary references has been created in the `.bib` format, these can be used to generate citations in LaTeX, as well as a reference section. In the following sections, we will cover the usage of the `biblatex` package. While `natbib`, which uses BibTeX as a backend, has some advantages such as creating customised bibliography style files using the makebst utility, and that numerous `.bst` files exist from journals that format the bibliography in the required style, it nevertheless is no longer being actively developed, and it is not as easy to modify existing styles as it is with `biblatex`. A guide to using `natbib` can be found at https://ctan.org/pkg/natbib.

### 2.1.1 Managing Bibliographies with BibLaTeX

To start using biblatex, we have to import the package in the preamble of the document, using `\usepackage{biblatex}`, as well as the bibliography `.bib` file, using `\addbibresource{references.bib}`, in this case importing a `.bib` file called references.bib (note that multiple `.bib` files can be imported by repeated use of this command). When loading the `biblatex` package, one can already specify how the citation and bibliography are to be formatted, with the `style` and `sorting` options respectively. In the example shown in Listing 2.4, we are using the citation shown in Listing 2.2. Note that if compiling from the terminal, you must first compile with pdflatex, then bibtex, and then again with pdflatex twice.

```
1   \documentclass{article}
2   \usepackage[
3   style=alphabetic,
4   sorting=nyt
5   ]{biblatex}
6
7   \addbibresource{references.bib}
8
9   \title{Sample document citing using the biblatex package}
10  \begin{document}
11
12  \noindent This is a sample sentence requiring a citation \cite{Smith2021}.
13
14  \printbibliography
15
16  \end{document}
```

**Listing 2.4** Example document using biblatex

This is a sample sentence requiring a citation [Smi21].

## References

[Smi21]   John Smith. *Formatting Bibliographies in LaTeX*. Springer, 2021.

**Fig. 2.2**  Output generated with code shown in Listing 2.4

This results in the document shown in Fig. 2.2.

### 2.1.1.1  Citation Options

Different citation styles exist for biblatex, with the `style` option in the `usepackage` command specifying how citations are to be printed. For most citation styles, a variety of citation commands exist in biblatex. The basic citation command is `\cite{}`, where the label of a particular reference in the `.bib` file is indicated in curly brackets. The citation is then formatted according to the global options given in the preamble, usually without any brackets if using `\cite{}`, except when the specified style specifies this (such as `alphabetic` or `numeric`). Nevertheless, citations can be formatted locally, with `\parencite{}` placing citations in round brackets, and `\footcite{}` placing citations in footnotes - if your chosen citation style allows for this! Some more specific citation commands are `\footcitetext{}`, which in addition to placing citations in the footnotes, also decreases the font size to `footnotesize`. Usually, when using `\textcite{}` the citation is integrated into the running text. These additional commands generally are used in conjunction with author-year styles in biblatex (e.g. `authoryear`), and tend not to be provided with numerical styles.

Many journals have their own biblatex styles, (such as `nature`, `science`, `ieee`) and other popular citation styles are also available (e.g. `chicago-authordate`, `mla`, `apa`). See for example, in Listing 2.5, the usage of the APA citation style.

```
1  \documentclass{article}
2  \usepackage[style=apa]{biblatex}
3
4  \addbibresource{references.bib}
5
6  \title{Sample document citing using the biblatex package}
7  \begin{document}
8
9  \noindent This is a sample sentence requiring a citation \parencite{Smith2021}.
10 As \textcite{JonesSmith2013} have shown, we can cite here too.
11
12 \printbibliography
13
14 \end{document}
```

**Listing 2.5**  Example document using biblatex

This results in the output shown in Fig. 2.3.

This is a sample sentence requiring a citation (Smith, 2021). As Jones and Smith (2013) have shown, we can cite here too.

## References

Jones, A., & Smith, W. (2013). A groundbreaking journal paper. *Journal Name*, *35*, 122–135.
Smith, J. (2021). *Formatting bibliographies in latex*. Springer.

**Fig. 2.3** An example of using the apa style with biblatex, see Listing 2.5

**Table 2.1** Citation commands for common biblatex styles

| Command | Citation |
|---|---|
| style=authoryear | |
| \cite{JonesSmith2013} | Jones and Smith 2013 |
| \cite[cf.][pages24-25]{JonesSmith2013} | cf. Jones and Smith 2013, pages 24-25 |
| \parencite{JonesSmith2013} | (Jones and Smith 2013) |
| \textcite{JonesSmith2013} | Jones and Smith (2013) |
| \cite*{JonesSmith2013} | 2013 |
| \parencite*{JonesSmith2013} | (2013) |
| style=numeric | |
| \cite{JonesSmith2013} | [1] |
| \cite[cf.][pages24-25]{JonesSmith2013} | [cf. 1, pages 24-25] |
| \supercite{JonesSmith2013} | [1] |
| \textcite{JonesSmith2013} | Jones and Smith [1] |
| style=alphabetic | |
| \cite{JonesSmith2013} | JS13 |
| \cite[cf.][pages24-25]{JonesSmith2013} | [cf. JS13, pages 24-25] |
| \supercite{JonesSmith2013} | [1] |
| \textcite{JonesSmith2013} | Jones and Smith [JS13] |

Citations can also be customised, with wording to appear before the citation, and after the citation, with \cite[before][after]{reference}. Note that the 'prenote' and 'postnote' are included in square brackets, while the bib entry key is in curly brackets. For three basic styles, we show the corresponding output in Table 2.1.

Other citation commands also exist, such as \fullcite{}, which prints the full reference and \nocite{}, which only serves to include a reference in the bibliography, without citation. To cite partial information from the reference, commands such as \citeauthor{}, \citetitle{}, or\citeyear{} exist.

### 2.1.1.2   Bibliography Formatting

While many biblatex styles already specify how the bibliography is to be generated, some standard styles such as alphabetic and numeric allow different sorting options to be chosen from. The default sorting option is nty, i.e. the references in the bibliography will be sorted first by name (author), then title, and then year. Other options can be specified in the usepackage command - for example, nyt, sorted by name, then year, then title, ydnt, sorted by descending year, name, and then title, or simply none, which sorts references by the order they appear in the text. Bibliographies can also be subdivided, for example to dedicate a separate bibliography section for each chapter, or to split the bibliography by the kind of source used, or a custom defined separation based on some keywords. To print all cited references that are journal articles, one can use the command \printbibliography{} with the option [type=article], and then the remaining cited references with the option [nottype=article], as shown in Listing 2.6.

```
1   \printbibheading
2   \printbibliography[type=article,heading=subbibliography,
3                               title={Journal Article Sources}]
4   \printbibliography[nottype=article,heading=subbibliography,
5                               title={Other Sources}]
```

**Listing 2.6**  Separating bibliographies by document type

The resulting output, when applied to our previous example, is shown in Fig. 2.4.

Bibliographies can also be printed based on a manually set keyword in the .bib file. If you include the field keywords={Keyword1,Keyword2,...} in a .bib entry, you can print a bibliography only for those references that are cited and that contain a particular keyword, as shown in the Listing 2.7

```
1   \printbibheading
2   \printbibliography[keyword=Keyword1,heading=subbibliography,
3                               title={Key References}]
```

**Listing 2.7**  Printing bibliography based on keyword

This is a sample sentence requiring a citation (Smith, 2021). As Jones and Smith (2013) have shown, we can cite here too.

## References

### Journal Article Sources

Jones, A., & Smith, W. (2013). A groundbreaking journal paper. *Journal Name*, *35*, 122–135.

### Other Sources

Smith, J. (2021). *Formatting bibliographies in latex*. Springer.

**Fig. 2.4**  Output when generating separate bibliographies, see Listing 2.6

One can also create bibliographies, for example, for each chapter in the book document class, by using the option `[refsegment=chapter]` when loading `biblatex`, and simply using `\printbibliography` `[segment=\therefsegment,heading=subbibliography]` at the end of each chapter. Using `[heading=subbibliography]` changes the heading of the bibliography to 'References', and places it immediately after the end of the text of the chapter (and not on a new page). The command `\printbibliography` can still be used at the end to print a bibliography of the entire document. The title of a bibliography can be easily changed by using the option `[title={}]` with the command `\printbibliography`. Alternatively, the headings for these segments can also be customised globally in the preamble of the document, for example by including the lines in Listing 2.8 at the beginning of the document.

```
1  \defbibheading{subbibliography}{%
2  \section*{Sources for Chapter \ref{refsegment:\therefsection\therefsegment}}}
```

**Listing 2.8** Printing custom bibliography title for each chapter

A `refsegment` can also be set manually, by using the environment `\begin{refsegment}` and `\end{refsegment}`. A bibliography can be then generated anywhere, with the command `\printbibliography` `[segment=1,heading=subbibliography]`.

### 2.1.1.3 Custom Styles

Depending on your requirements, it is often easiest to find a bibliography style that is closest to your preferences, and then tweak it based on the various options you can pass to the biblatex package when loading it. Some common options include `[maxnames={}]`, which sets the maximum number of authors to show in citations and bibliography (they can also be set individually with options `maxbibnames` and `maxcitenames`); `url`, `isbn`, `doi`, can be used as options, and set equal to either `True` or `False`, which can ignore these entries that are often included in automatically generated .bib files. The option `[giveninits=true]` will print all first or middle names as initials only.

## 2.2 Editors

To edit a latex document one only requires a text editor to write a document - this can then be compiled with LATEX from the command line. Nevertheless, there is a myriad of editors out there facilitating this process, with many features to help typeset documents. While offline editors have long been the most commonly used option, a number of online editors have become popular in recent years.

### 2.2.1  Offline Editors

While many starting out with LaTeX nowadays start off with online editors for the sake of simplicity in setting them up, there are some considerable advantages to installing an offline LaTeX editor. For one, you might not always have an internet connection and need to edit a document. Then, there is the compilation time factor for larger documents (some free online editors have compile timeout limits), and, most offline editors allow the usage of customised key combinations for commands which can be useful. Last, but not least, if one already uses an editor to code in other programming languages, it might be easiest to use a familiar environment for typing LaTeX documents.

Many popular offline LaTeX editors exist, with the most popular offline editors listed in Table 2.2.

All of these are free, and most of them have included PDF viewers allowing you to preview the final document. Some include code completion, check spelling, can help with file management and version control, among other feature facilitating the use of LaTeX.

As mentioned, only a text editor is technically required for typesetting a LaTeX document. Hence, any editor used for programming can also be used to edit and compile LaTeX documents. For example, Atom (WLM), Notepad++ (W), Sublime Text (WLM), Emacs or g/n/vim (WLM) can be used for this purpose. For vim, some plugins may make the whole LaTeX typesetting experience a lot more efficient and user-friendly, by allowing commands/mappings for compilation, PDF previewing, forward and reverse search, command completion, reference completion, better syntax highlighting, document navigation, and obviously the many myriad features and power of using vim as a text editor. There are currently two main options, namely `Latex-Suite`/`vim-latex` and `vimtex`. A guide for `Latex-Suite`/`vim-latex` can be found at http://vim-latex.sourceforge.net/documentation/latex-suite-quickstart.html, and for `vimtex`, which we recommend, at https://github.com/lervag/vimtex.

**Table 2.2**  List of most popular LaTeX editors. OS key: W/indows, L/inux, M/ac OS

| Editor | OS support | URL |
|---|---|---|
| TeXstudio | WLM | www.texstudio.org |
| TeXshop | M | https://pages.uoregon.edu/koch/texshop/ |
| TeXnicCenter | W | www.texniccenter.org |
| TeXmaker | WLM | www.xm1math.net/texmaker/ |
| TeXworks | WLM | www.tug.org/texworks/ |

### 2.2.2   Online Editors

Online LaTeX editors have a number of advantages, that may make these more attractive to you than installing your own offline distribution. For one, there is no need to install any LaTeX distribution, or worry about any updating or installation issues. Additionally, there is the convenience factor, of having your documents accessible from the cloud at any point in time. For beginner LaTeX users, this may be very convenient, as well as the many templates often available on such online editors.

There are, nevertheless, some disadvantages to highlight. First of all, there are the conditions that such services set - for example, many free accounts may come with limitations, with regards to upload limits (e.g. of figures), compile time limits (thus only allowing compilation of smaller documents), which may only be overcome by paying for a subscription. Then, one must consider any security and reliability requirements - as with any other cloud internet service, there may be breaches of security, server failures leading to lost documents, and server downtime required for maintenance. For more advanced users, being tied to a particular editor, and a particular distribution of LaTeX and not having full control may prove a nuisance.

Regardless, online LaTeX editors are becoming increasingly popular, and are particularly useful for quicker and shorter documents. The most popular services currently available are www.overleaf.com, www.cocalc.com, and www.papeeria.com. Many universities are also offering paid-level feature access for some of these, which may be worth looking into if one is interested in these services.

## 2.3   Page Setup

While the size of the page in LaTeX is set in the beginning with the `\documentclass` command (for example, with `\documentclass[a4paper]{article}`), LaTeX allows for much more fine-tuning of how the space on the page is used. Two packages are particularly useful for this, namely the `geometry` package and the `fancyhdr` package.

First we must briefly review the spaces that are available in LaTeX, as many sizes in LaTeX can be given relative to these. The total height and width of a page are given as `\pageheight` and `\pagewidth`, while the text area of the page is given by `\textwidth` and `\textheight`. The width of an actual line of text is given by `\linewidth`, and the distance between lines in a paragraph is given by `\baselineskip`, while the distance between paragraphs is given by `\parskip`. These measures are useful to know, as one can easily define the sizes of objects on the document page relative to these. For example, a figure can be scaled to have a width equal to 50% of the line width, with the command `\includegraphics[width=0.5\linewidth]{}`. Alternatively, LaTeX also recognises absolute measures, in cm, mm, pt, sp, in, and em/ex (the heights of an 'M' and 'x', respectively).

### 2.3.1  Margins and Page Size

To manually set margins in a LaTeX document, we can simply use the geometry package as in Listing 2.9:

```
\usepackage[top=1cm, bottom=1cm, right=2.54cm, left=2.54cm]{geometry}
```
**Listing 2.9**  Setting margins with the geometry package

We can also use the `geometry` package to specify the paper size and the text area size, with the option `total`. For example with the example shown in Listing 2.10, the page will be in standard A4 size, with the text occupying an area of 16 cm by 21 cm.

```
\usepackage[a4paper, total={16cm, 21cm}]{geometry}
```
**Listing 2.10**  Setting the paper size and the text area size with the geometry package

The standard paper sizes available are: `a0paper` to `a6paper`, `b0paper` to `b6paper`, `letterpaper`, `legalpaper`, and `executivepaper`. One can also define custom paper sizes, by including the option [`paperwidth=`] and [`paperheight=`]. In general then, `paperwidth` is equal to the lengths `left`, `width`, and `right`, while the length `paperheight` is equal to `top`, `height`, and `bottom`. In defining the size of the page, by default the size of the header and footer are excluded - the height of the header is defined by `headheight`, the distance of the header from the text by `headsep` and the distance from the text to the bottom of the footer by `footskip`, and its height by `footheight`. One can also include the heights of the header and footer in the total text area, by passing the options `includehead`, `includefoot`, or `includeheadfoot`. Thus, if one desired to have a header that goes all the way to the top of the page, and a footer that goes all the way to the bottom of the page, one can use the example shown in Listing 2.11, where the top and bottom margins are set to 0 cm, and the header and footer are made to be include in the text page area, and their dimensions are set.

```
\usepackage[top=0.0cm, bottom=0.0cm, right=2.54cm, left=2.54cm,
includeheadfoot, headheight=2.54cm, footskip=2.9cm
]{geometry}
```
**Listing 2.11**  Setting header and footer sizes with margins

Now this example might seem a bit strange - why would one want to do this? One example where such a function might come in useful, is if one wanted to have a banner at the top and bottom of the page, with no white space in between the margins and the banner, which we will see shortly.

### 2.3.2  Colors

There is a standard set of colors in LaTeX (white, black, blue, yellow ...) that can be used with the package `xcolor` or `color`, see Table A.2 for more options. The package `xcolor` can also be included with the option `dvipsnames`, allowing a larger set of predefined colors to be accessed. Furthermore, custom colors can be defined, using the command `\definecolor{name}{format}{color}`, where `format` defines the input type, such as hexadecimal input HTML, hue/saturation/brightness, or cyan/magenta/yellow/k. Colors can also be mixed - for example, to achieve a color that is 30% green and 70% red, one can use `\colorlet{myColor}{green!30!red!70!}`, and its complimentary color by `\colorlet{compMyColor}{-green!30!red!70!}`. One can also define colors relative to other colors, for example, to have a color that has 60% of the intensity of a previously defined colors, we can use `\colorlet{newColor}{myColor!60!}`. This can be useful for setting some text in a different in color using the command `\textcolor{color}{text}`, while the background of some text can be defined using the command `\colorbox{color}{text}`. We can also modify the page color with the command `\pagecolor{color}` (and revert with `\nopagecolor`). To set the color of the text for the entire document, one can (define a color and) simply use `\color{}` in the preamble. One can also use `\color{}` at a particular point in the document, changing the color of the text from that point on. Colors are useful too when creating (text)boxes, using the commands `\colorbox{color}{text}` and `\fcolorbox{frame color}{background color}{text}`, where the former is without a frame, and the latter allows one to specify a (colored) frame.

### 2.3.3  Headers and Footers

To have more elaborate headers and footers, the package `fancyhdr` is very useful. While, for example, the `book` documentclass provides some headers throughout the document, `fancyhdr` allows for full customisation of the header in your documents. Headers and footers are set in the preamble, as shown in the Listing 2.12. The simplest case is specifying what should be printed in the right and left corners of the header and footer, as well as centered fields, with the commands `\rhead{}`, `\lhead{}`, `\chead{}`, `\rfoot{}`, `\lfoot{}`, and `\cfoot{}`. Some special fields can be included, such as `\thepage` which will generate the page number, and similarly `\thechapter` and `\thesection`, while `\leftmark` and `\rightmark` generate the name and number of the current Chapter/Section for (reports or book/article), and Section/Subsection (for reports or book/article), respectively. A simple example is shown in Listing 2.12.

```
1  \usepackage{fancyhdr}
2    \pagestyle{fancy}
3    \fancyhf{}
4    \rhead{Report Title}
5    \lhead{Author}
6    \rfoot{\thepage}
7    \lfoot{2021}
```

**Listing 2.12**  Setting header and footer with fancyhdr

Particularly useful for two-sided documents, one can specify where exactly each of these elements should be located using the command \fancyhead{} and \fancyfoot{} in the option square brackets, with L and R indicating left and right pages, and O and E denoting odd and even pages, and C indicating a centered object, as shown in Listing 2.13.

```
1   \usepackage{fancyhdr}
2     \pagestyle{fancy}
3     \fancyhf{}
4     \fancyhead[LE,RO]{Report Title}
5     \fancyhead[RE,LO]{Author}
6     \fancyfoot[LE,RO]{\thepage}
7     \fancyfoot[CE,CO]{2021}
```

**Listing 2.13** Customising headers and footers in two-sided documents with fancyhdr

The thickness of the lines separating the header and footer from the text can be specified using the commands shown in Listing 2.14. Setting 0pt for any line will remove the line.

```
1   \renewcommand{\headrulewidth}{0pt}
2   \renewcommand{\footrulewidth}{2pt}
```

**Listing 2.14** Defining lines for header or footer with fancyhdr

In the following example we will attempt to generate a banner for the top and bottom of our page, to help exemplify some of these commands. Thus, in the preamble, we will begin by eliminating the lines between header, footer, and the text, and make the header cover our right and left margins as well. To do this, we use the command \fancyhfoffset[]{}, and set it equal to our right and left margins, in this example, 2.54 cm, thus extending our header and footer to the edges of the document, as shown in Listing 2.15. Here, the packages xcolor and graphicx are also called, to allow for color selection and figure support.

```
1    \documentclass{article}
2    \usepackage[top=0.0cm, bottom=0.0cm, right=2.54cm, left=2.54cm,
3    includeheadfoot, headheight=2.54cm, footskip=2.9cm
4    ]{geometry}
5    \usepackage{graphicx,fancyhdr}
6    \usepackage[dvipsnames]{xcolor}
7    \fancyhf{}
8    \pagestyle{fancy}
9
10   \renewcommand{\headrulewidth}{0pt}
11   \renewcommand{\footrulewidth}{0pt}
12
13   \fancyhfoffset[L]{2.54cm}
14   \fancyhfoffset[R]{2.54cm}
```

**Listing 2.15** Setting margins for header and footer extending to page edge

To then include a fully colored, rectangular banner across the top of the page, we can use the \fancyhead{} command, and centre a box within it. To get boxes in LATEX, one can use the command \fbox{}, but if we want a colored box, we can use \fcolorbox{frame color}{box background color}{...}. We can use any of the predefined colors (most common colors - black, green, blue, red ...) or define our own color using the xcolor package. To assign text or images to different

positions in our header and footer, we will use fractions of the total horizontal length of the header \headwidth, using several minipages. Note the options for the minipage environment, which allows setting the alignment of both the box, and the content within the box, as well as heights and widths. The widths here will be set relative to the headerwidth, and the height to be about the size of the top margin of the paper (i.e. the vertical height of our header and footer). The syntax for minipage is
\begin{minipage}[alignment of box][height][alignment of content in minipage]{width of box} ... \end{minipage}%

Note also that defined spaces between elements of the header are easiest added with an empty minipage, as we are using fractions of headerwidth, as shown in Listing 2.16; furthermore, we have used different font colors and text sizes. In this example, a sample image with the name 'logo.jpg' is located in the same folder as the document.

```
1  \fancyhead[C]{%
2  \noindent%
3  \fcolorbox{SpringGreen}{SpringGreen}{%
4  \begin{minipage}[c][2.3cm][c]{0.025\headwidth}
5  \mbox{}
6  \end{minipage}%
7  \begin{minipage}[c][2.3cm][c]{0.5\headwidth}
8  \LARGE{\textcolor{darkgray}{This is the Title of a Very Important Report}}\\\large{\textcolor{gray}{Author}}
9  \end{minipage}%
10 \begin{minipage}[c][2.3cm][c]{0.45\headwidth}%
11 \hfill%
12 \includegraphics[height=0.9\headheight]{logo.jpg}
13 \end{minipage}%
14 \begin{minipage}[c][2.3cm][c]{0.025\headwidth}
15 \mbox{}
16 \end{minipage}%
17 }}
```

**Listing 2.16**  Settings for fancyhdr to generate top page banner

Similarly now, for the footer, we will repeat these settings, as shown in Listing 2.17. Note that the page number is being called using the command \thepage, and that it is aligned centrally in the box, rather than the left and right text fields, which we have aligned to the bottom of the minipage. In both left and right fields, we have added an empty line under the text, such that the actual text is not entirely at the bottom of the page; alternatively, this could be moved with the command \vspace{}.

```
1  \fancyfoot[C]{%
2  \noindent%
3  \fcolorbox{SpringGreen}{SpringGreen}{%
4  \begin{minipage}[b][2.54cm][c]{0.025\headwidth}
5  \mbox{}
6  \end{minipage}%
7  \begin{minipage}[b][2.54cm][b]{0.45\headwidth}%
8  {\textcolor{gray}{Organisation Title}}\\\mbox{}
9  \end{minipage}%
10 \begin{minipage}[b][2.54cm][c]{0.05\headwidth}
11 \thepage
12 \end{minipage}%
13 \begin{minipage}[b][2.54cm][b]{0.45\headwidth}
14 \flushright{\textcolor{gray}{www.url.co.uk}}\\\mbox{}
15 \end{minipage}
```

```
16   \begin{minipage}[b][2.54cm][c]{0.025\headwidth}
17   \mbox{}
18   \end{minipage}%
19   }}
```

**Listing 2.17**   Settings for fancyhdr to generate bottom page banner

This results in the banners shown in Figs. 2.5 and 2.6.

## 2.4   Converting to and from LaTeX

Sometimes it might be necessary to convert a LATEX document into a different format, such as .docx. A number of tools - free and paid - exist that are able to perform this. The tool Writer2LaTeX, available at http://writer2latex.sourceforge.net/, allows for the conversion of OpenDocument (.odf) documents to LATEX. The tool rtf2latex2e converts rich text format (.rtf) files to LATEX (runs on MS Windows, macOS and unix). MS Word files can be saved as both .rtf or .odf files. Some paid tools exist too - such as the online page Docx2LaTeX, at https://www.docx2latex.com/, a service allowing conversion of .docx documents to LATEX. Furthermore, GridEQ has two shareware tools 'Word-to-LATEX' and 'LATEX-to-Word', that allow conversion between MS Word and LATEX. It only runs on MS Windows, and has a limited number of free runs. Other workarounds include attempting to convert from the PDF, for example to generate HTML from PDF.

Converting to MS Word may be the most useful use-case, as sometimes journals may require submission in .docx or .odf. There are two main ways of doing this. First, one can attempt to open the .pdf file in a text editor like MS Word. The other solution is to use the `pandoc` tool, which can be downloaded form www.pandoc.org. Pandoc is used from the command line, as shown in the first line of the Listing below. Pandoc can handle a number of formats, for example one can convert LATEX to markdown, using the command shown in the second line in Listing 2.18.

**Fig. 2.5**   Result of the header as generated by fancyhdr from the example code in Listing 2.16

**Fig. 2.6**   Result of the footer as generated by fancyhdr from the example code in Listing 2.17

```
1  pandoc document.tex –o document.docx
2  pandoc document.tex –o document.md
```

**Listing 2.18**  Using pandoc to convert .tex files

The conversion is obviously not perfect, and may require some manual changes. Nevertheless, equations are usually well typeset, and bibliographies with citations can also be set, using the option `--bibliography`, and cross-referencing with the tool `pandoc-crossref` (available at https://github.com/lierdakil/pandoc-crossref), as shown in Listing 2.19.

```
1  pandoc document.tex –filter pandoc–crossref –bibliography=references.bib –o document.docx
```

**Listing 2.19**  Bibliography conversion using pandoc

LaTeX output can also be generated from other programs, such as Python or R. In Python, for example, the `tabulate` package can be used to generate tables in LaTeX code directly, for example to print the numpy array `myTable` with a defined list of headers, as LaTeX output, one can use `print(tabulate(myTable, headers, tablefmt="latex"))`, after importing the package `tabulate` in Python.

Graphs generated in Python, for example with `matplotlib`, can also be generated for easy inclusion in LaTeX documents by setting the typical Computer Modern LaTeX font, setting the option as shown in Listing 2.20, and then saving the plot as a .eps or .pdf file.

```
1  import matplotlib.pyplot as plt
2  plt.rcParams.update({
3      "text.usetex": True,
4  })
5  plt.savefig('plot.pdf')
```

**Listing 2.20**  Setting matplotlib font in Python

Other fonts can also be included and specified. Alternatively, one can also use the package `pgf` in LaTeX to include a graph generated in Python, in such a way that is generated in LaTeX. This has the advantage, that the same font is used as in the document. The resulting .pfg file can then simply be included using the command `\input{plot.pgf}` in a `figure` environment, as shown in Listing 2.21.

```
1  import matplotlib as mpl
2  mpl.use("pgf")
3  import matplotlib.pyplot as plt
4  plt.rcParams.update({
5      "text.usetex": True,
6  })
7  ...
8  plt.savefig('plot.pgf')
```

**Listing 2.21**  Generating .pgf file with matplotlib

## 2.5   XeLaTeX and Font Specification

Technically, LATEX is simply a collections of macros for TEX - for example, the commands \section{} or \usepackage{} are defined herein. As TEX itself only takes 8 bit characters as inputs, and produces .dvi (DeVice independent format) files, several adaptations of TEX have been created to help facilitate the usage of TEX, often referred to as 'TEX engines'. The most common TEX engines are pdf-TeX, XeTeX, and LuaTeX. The most commonly used TEX engine today is pdfTeX, incorporating the e-TeX extensions, that directly generates a .pdf file from the .tex input file. A major benefit of this is the capacity to add PDF functionality to documents (hyperlinks, document information, among others...), notably via the use of the hyperref package, and has some added microtypography improvements (kerning options, among others...), such as those found in the microtype package. It is included with most common LATEX distributions, such as MiKTeX or TeX Live. Prior to pdfTeX, (La)TeX output was commonly first converted to postscript, and only then converted to PDF as an additional step.

### 2.5.1   XeLaTeX

XeLaTeX has the ability to directly read UTF-8 encoded unicode text and additional font formats such as OpenType (OTF) fonts. This means that we can select different fonts with quite a lot of ease. For example, to set the EBGaramond font in our main document from a .otf file, we can use the code as shown in Listing 2.22, using the command \setmainfont[]{}, provided by the package fontspec - note that here, we are providing LATEX with the exact font files in the .ttf/.otf format. Generally, it is not necessary to specify separately the italic or bold fonts, but we have done so here to show how one can do this manually - in this case to change the default bold EBGaramond font to the medium version of the font.

```
1   \usepackage{fontspec}
2   \setmainfont[ItalicFont={EBGaramond-Italic},
3       BoldFont={EBGaramond-Medium},
4       BoldItalicFont={EBGaramond-MediumItalic}
5       ]{EBGaramond-Regular.otf}
```

**Listing 2.22**   Setting specific fonts in XeLaTeX

Similarly, one can set different types of fonts, for example for verbatim environments with the command \setmonofont{}. XeLaTeX is the most flexible when it comes to including and selecting fonts, but nevertheless, fonts can also be changed in pdfLaTeX.

## *2.5.2   Fonts in LaTeX Generally*

In regular pdfLaTeX, we can also select specific fonts for different functions. For example, we can change the font for a particular section by placing text in a `{\fontfamily{fontcode}\selectfont ... }` environment. Fonts can be loaded usually via packages in pdfLaTeX, and then have a particular font code that is used when specifying a font. For example, the helvetica font is included with `\usepackage{helvet}`, and can be used with the font code phv.

One can also change the default fonts for particular document elements, such as section or subsection headings, particularly with the package `sectsty`. For example, to change the size and font of the section and subsection headings, we can use the `\sectionfont{}` and `\subsectionfont{}` commands, as shown in Listing 2.23.

```
1  \sectionfont{\LARGE\fontfamily{phv}}
2  \subsectionfont{\Large\fontfamily{phv}}
```

**Listing 2.23**   Font specification in LaTeX

We can also include more complex options, such as modifying subsubsection headings, as well as the paragraph fonts, and change headings to be entirely in uppercase, as shown in Listing 2.24.

```
1  \sectionfont{\Large\fontfamily{phv}\selectfont\uppercase}
2  \subsectionfont{\large\fontfamily{phv}\selectfont\uppercase}
3  \subsubsectionfont{\normalsize\fontfamily{phv}\selectfont}
4  \paragraphfont{\normalsize\textit}
```

**Listing 2.24**   More complex font specifications

To customise our section headings even more, we can also for example change the numbering to Roman numerals, as shown in Listing 2.25.

```
1  \renewcommand{\thesection}{\Roman{section}.}
2  \renewcommand{\thesubsection}{\Roman{section}.\Roman{subsection}.}
3  \renewcommand{\thesubsubsection}{\roman{subsubsection}.}
```

**Listing 2.25**   Roman numerals in section headings

## 2.6   Additional Environments

## *2.6.1   Text Formatting*

### 2.6.1.1   Languages

In LaTeX, it is fairly easy to switch between language inputs - most of the time, it is sufficient to include `\usepackage[utf8]{inputenc}` in the preamble, and any utf8-encoded font can be directly given as input to LaTeX. Nevertheless, it is possible for LaTeX to transliterate certain scripts. For example, with the `babel` package, one can include support for a number of languages. `babel` can, and

**Table 2.3**  Common special characters in LaTeX

| grave accent | \`{o} | ò |
|---|---|---|
| acute accent | \'{o} | ó |
| circumflex | \^{o} | ô |
| umlaut, trema or dieresis | \"{o} | ö |
| long Hungarian umlaut (double acute) | \H{o} | ő |
| tilde | \~{o} | õ |
| cedilla | \c{c} | ç |

should, generally be loaded in any case, as it gives LaTeX the correct hyphenation rules, which vary even between UK and US English - and can be loaded with `\usepackage[USenglish]{babel}` or `\usepackage[UKenglish]{babel}`. Multiple languages can also be selected, by adding more languages when loading the package, with the last language indicated serving as the default language, for example with

`\usepackage[greek,russian,UKenglish]{babel}`.

To switch between these languages, one can use the command `\selectlanguage{language}`, and then use it again to revert to the default language. Alternatively, one can also use `\foreignlanguage{language}{text}`, to typeset shorter pieces of text in a different language.

One can also typeset languages in different scripts, which can sometimes prove slightly challenging. For example, directly typing in Greek script requires both `\usepackage[utf8]{inputenc}` for the input, but also `\usepackage[LGR]{fontenc}` to correctly produce Greek letters. When combining scripts, for example both Greek and Latin scripts, one can use:

```
1  \usepackage[utf8]{inputenc}
2  \usepackage[LGR,T1]{fontenc}
3  \usepackage[greek,UKenglish]{babel}
```

**Listing 2.26**  Example of combining Greek and Latin scripts

Some basic special characters can be generated using the commands shown in Table 2.3.

Some special cases require special packages - for example, typesetting text in ancient Greek script, where letters can contain multiple accents. Here, the option `polutonikogreek` can be passed to the `babel` package, to transliterate the text as shown in the Listing 2.27 below.

```
1  \usepackage[polutonikogreek, UKenglish]{babel}
2  \selectlanguage{polutonikogreek}
3  alpha\\
4  \selectlanguage{UKenglish}
5  alpha\\
```

**Listing 2.27**  Ancient Greek script with babel package

**Table 2.4**  Package options to load various scripts in pdfLaTeX

| Arabic | \ usepackage{arabtex} |
|---|---|
|  | \usepackage[utf8]{inputenc} |
|  | \usepackage[LFE,LAE]{fontenc} |
|  | \usepackage[arabic]{babel} |
| Chinese | \usepackage{CJKutf8} |
|  | \begin{CJK*}{UTF8}{gbsn} |
|  | \end{CJK*} |
| Japanese | \usepackage{CJKutf8} |
|  | \begin{CJK*}{UTF8}{min} |
|  | \end{CJK*} |
| Korean | \usepackage{CJKutf8} |
|  | \begin{CJK}{UTF8}{} |
|  | \CJKfamily |
|  | {mj} |
|  | \end{CJK} |
| Russian | \usepackage[T2A]{fontenc} |
|  | \usepackage[utf8]{inputenc} |
|  | \usepackage[russian]{babel} |

which will produce the words αλπηα and alpha. Another option, for this particular case, is considering the package `betababel`.

Of course, as long as one can type the words in the document, there is technically no need for this, at least for utf-8 based text. Hence, we can type 'ὅπερ ἔδει δεῖξαι', with only the `fontenc` and `inputenc` packages.

For other languages, similar packages exist that provide the right encoding support or fonts, and are shown in Table 2.4.

#### 2.6.1.2  Verbatim

To display code in LaTeX, the most commonly used environment is `verbatim`, highlighting the code text by changing the font. It can be used by simply using the environment `\begin{verbatim}` ... `\end{verbatim}` Here, any commands used are helpfully ignored, and code on new lines is actually printed on new lines. Verbatim text can also be used in-line, by using the `\verb! ... code ... !`, where the exclamation marks can be substituted by any symbol (but *), to delimit the verbatim text. The package `lstlisting` allows for even better displaying of code, with code highlighting and better management of spaces. For this, one can also specify the coding language being used. A basic example for a `lstlisting` environment is shown in Listing 2.28.

```
1  \begin{lstlisting}[language=Python]
2      # This program prints 'Hello, world!'
3      print('Hello, world!')
4  \ end{lstlisting}
```

**Listing 2.28**  Usage of the lstistings package

A nice feature of lstlisting is that code can also be included from a file by simply using, in this case, the command `\lstinputlisting[language=Python]{printEx1.py}`. Sections of code can be specified, by using the options `[firstline=]`, and `[lastline=]`, and a caption can be set with the option `[caption =]`, and labels with `[label=]`, which can then be used with `\ref{}` to cross-reference the listing.

### 2.6.1.3  Linenumbers

The `lineno` package provides the `linenumbers` environment, that allows for the setting of text blocks with numbered lines. This is very useful when setting text extracts or even primary sources in LATEX. An example is shown in Listing 2.29, where instead of the regular `\begin{linenumbers}[n]` and `end{linenumbers}` commands (where *n* indicates the linenumber to start from), the usage of it within a minipage is shown, which requires the usage of `\begin{internallinenumbers}`. Various options exist, one of the most useful being the option `\modulolinenumbers[n]`, which allows one to only print every *n*-th line.

```
1   \begin{minipage}{0.5\textwidth}
2   \noindent \textbf{you are welcome to elsinore}\\
3
4   \modulolinenumbers[2]
5   \begin{internallinenumbers}[1]
6
7   Between us and the words there is molten metal
8
9   between us and the words there are helices that run
10
11  and can give us death \,\,\, ravish us \,\,\, remove
12
13  from the most deep of us the most useful secret
14
15  ...
16
17  \end{internallinenumbers}
18  \end{minipage}%
19  \begin{minipage}{0.5\textwidth}%
20  \noindent \textbf{you are welcome to elsinore}\\
21
22  \modulolinenumbers[2]
23  \begin{internallinenumbers}[1]
24
25  Entre n\'{o}s e as palavras h\'{a} metal fundente
26
27  entre n s e as palavras h\'{a} h\'{e}lices que andam
28
29  e podem dar-nos morte \,\,\,\, violar-nos \,\,\,\, tirar
30
31  do mais fundo de n\'{o}s o mais \'{u}til segredo
32
33  ...
34
35  \end{internallinenumbers}
36  \end{minipage}
```

**Listing 2.29**  Usage of the linenumbers package

The result of which is shown in Fig. 2.7.

you are welcome to elsinore                          you are welcome to elsinore

Between us and the words there is molten metal       Entre nós e as palavras há metal fundente
₂ between us and the words there are helices that run ₂ entre nós e as palavras há hélices que andam
  and can give us death   ravish us   remove            e podem dar-nos morte   violar-nos   tirar
₄ from the most deep of us the most useful secret    ₄ do mais fundo de nós o mais útil segredo

**Fig. 2.7**  Example of the linenumbers environment, see Listing 2.29

## *2.6.2  LaTeX Macros*

### 2.6.2.1  Commands

Since LaTeX is just a collection of macros for TeX, we can obviously write our own macros - essentially commands to help us simplify our LaTeX workflow. To define a new macro in LaTeX, one can simply use the command
`\newcommand{\command}{definition}`, or
`\newcommand\command{definition}`.

In this first example, essentially whenever the `\command` is used, it is substituted in the compiled document by its definition. Thus, using Listing 2.30, the compiled document will read 'This book is called "An Introduction to LaTeX for Research and Education".

```
1   \newcommand{\title}{An Introduction to \LaTeX{} for Research and Education}
2   This book is called "\title".
```

**Listing 2.30**  Simple definition of new commands

`\newcommand` can be extended to take arguments as inputs, in which case the number of arguments needs to specified in the form
`\newcommand\command[number of arguments]{definition}`
where the argument's position is indicated by a hashtag (e.g. #1, #2...) For example, with Listing 2.31, we can print some simple text whenever we want to, changing the date by supplying it as arguments to the command. In this case, the listing below will print the statement "This report was published on January 1, in the year 2000."

```
1   \newcommand{\monthdayyear}[3]{This report was published on #1 #2,
2   in the year #3.}}
3   \monthdayyear{January}{1}{2000}
```

**Listing 2.31**  New commands with arguments

This may not seem *that* useful. But we must note, that even at such a simple level, text that requires more complex formatting can easily be generated whenever needed. For example, if we require that family names be rendered in bold font, we could define a command as shown in Listing 2.32, where 'Smith' will be printed in bold font. The tilde prevents the given and family names from being separated by a line break.

```
1  \newcommand\name[2]{#1~\textbf{#2}}
2  \name{John}{Smith}
```

**Listing 2.32**  Font formatting with newcommand

For each additional argument, a new set of curly brackets is added. Last, we should note that one can also set default values to be used, in the case no argument is explicitly supplied. In that case, we define a new command with
`\newcommand{name}[num][default]{definition}`.
Here, the default value will be used for the first argument, unless a different first parameter is given, as shown in Listing 2.33.

```
1  \newcommand\accessdate[3][2021]{This website was last accessed in the
2          year #1, on #2 #3.}
3  \accessdate{March}{1}
4  [...]
5  \accessdate{2020}{December}{30}
```

**Listing 2.33**  Setting default values for new commands

Generally, it is good practice to actually use the command `\newcommand*{}{}`, as this does not allow paragraph breaks or empty lines as arguments for commands.

#### 2.6.2.2  Environments

Similarly to how we have defined new commands, we can also define new environments, using the by now familiar `\begin{}` ... `\end{}` syntax we have seen for figure, tables, among others. The syntax required to achieve this is quite similar to `\newcommand{}`, using

> `\newenvironment{name}{before}{after}`

The content in the `before` brackets is executed once the `\begin{environmentName}` command is encountered; then, the contents between `\begin{environmentName}` and `\end{environmentName}` are executed, before the contents of the `after` brackets are executed. A simple example is given in Listing 2.34.

```
1  \newenvironment{largeQuote}{\begin{quote}\LARGE}{\end{quote}}
2  \begin{largeQuote}
3  This is a quote typeset in a LARGE font size.
4  \end{largeQuote}
```

**Listing 2.34**  Defining new environments

This code will render:

# This is a quote typeset in a LARGE font size.

### *2.6.3  Formatting Files*

If one is generating many documents where the same kind of formatting is required (letters, receipts, reports...), part of the work can be simplified by setting all the formatting in a separate text document, and simply including this file in the preamble when required, with \input{formatting.tex}. Further, one might desire to define settings for the formatting of the title, author, or any other field of the document, and keep these settings in a separate file to include, and not have to edit this file to enter the, for example, author, title, or whichever field one requires. The solution to this would be to use variables in the 'formatting document', and then define these in your actual document.

For example, we can define the style for the beginning of an automatically generated report, by setting it up as shown in Listing 2.35 by defining a command (here) called \header, creating variables for those elements that have to be defined for each document. This document can then be saved as formatting.tex.

```
1  \newcommand{\header}{
2  \noindent\textbf{\LARGE{\reportTitle}}\\\noindent\rule{\textwidth}{0.4mm}\\
3  \begin{minipage}{0.6\linewidth}
4  \begin{flushleft}
5  \reportSubtitle
6  \end{flushleft}
7  \end{minipage}%
8  \begin{minipage}{0.4\linewidth}
9  \begin{flushright}
10 \scriptsize{This report was made for\\ \customerName{} on\\ \reportDate.\\
11 This report was authorised by\\ \reportAuthor.}
12 \end{flushright}
13 \end{minipage}
14 \vspace*{1.5cm}
15 }
```

**Listing 2.35**  Using commands as variables

We can then set the specifics in the actual document, by defining each variable using \newcommand, and after this, the formatting file can be called, as shown in Listing 2.36.

```
1  \documentclass{article}
2  \newcommand{\reportAuthor}{James Smith}
3  \newcommand{\customerName}{Alice Richards}
4  \newcommand{\reportDate}{March 12, 2020}
5  \newcommand{\reportTitle}{Draft Title}
6  \newcommand{\reportSubtitle}{Draft Subtitle}
7
8  \input{formatting}
9  \begin{document}
10 \header\\
11 ...
12 \end{document}
```

**Listing 2.36**  Document using separate formatting file

Which outputs a header as shown in Fig. 2.8.

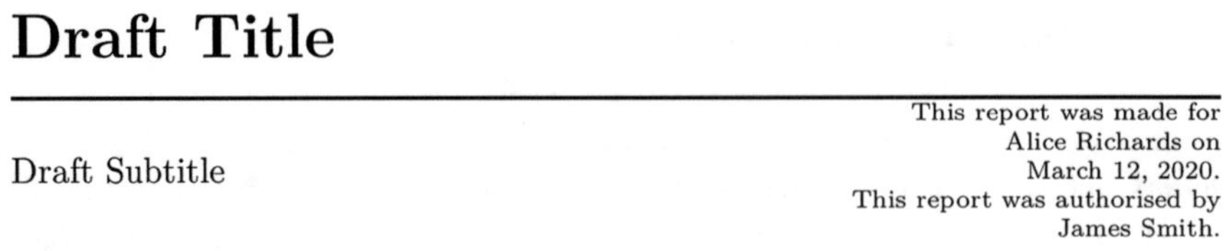

# Draft Title

Draft Subtitle

This report was made for
Alice Richards on
March 12, 2020.
This report was authorised by
James Smith.

**Fig. 2.8** Header generated with external file formatting, see Listings 2.35 and 2.36

## 2.6.4  Serial Documents

In some instances, it may be required to generate a larger number of documents, which have the same formatting. This may be useful for generating serial letters or report cards. Several approaches exist here - one could, similarly as in the previous section, simply define a command that prints the letter text and substitutes the necessary fields with the help of arguments. But what if we need more control, and to access more complex data to substitute in, for example tables or plots? For such purposes, it might be more useful to resort to a scripting language like Python.

A simple example would be that of a serial letter, where the required information (name, address, etc...) is provided in a .csv file (common output for spreadsheet files or databases). One solution would be to have Python write the entire LaTeX file, line by line - but this could be a bit cumbersome. A simpler approach consists of preparing your LaTeX file with commands as placeholders for the text you wish to substitute it with. For example, in the following Listing 2.37, two commands for the name and address of an individual are introduced. Note that they are defined, first, by some unique key (in the example, #letterName and #letterAddress). These will then be replaced by the Python script, as shown in Listing 2.38, for each letter individually from the .csv file. It is easiest to create one new folder for each series of documents, and place the template LaTeX file and Python script into it, and running the commands using the terminal.

```
1   \documentclass{letter}
2   \usepackage{lipsum}
3   \newcommand{\lname}{#letterName}
4   \newcommand{\laddress}{#letterAddress}
5
6   \address{\laddress}
7
8
9   \begin{document}
10  \pagenumbering{gobble}
11  \begin{letter}{Name of Institution\\ Street\\City}
12
13  \opening{Dear \lname ,}
14
15  \lipsum[1]
16  \end{letter}
17  \end{document}
```

**Listing 2.37** LaTeX template file for serial letter

As you can see in Listing 2.37, whenever we required the name of the individual being addressed, we can use \lname, and so on. In the Python code below, we first read in the .csv file which contains the names and addresses, shown in List-

10 Main Avenue

August 22, 2021

Name of Institution
Street
City

Dear Ms. Jane Doe,

Lorem ipsum dolor sit amet, consectetuer adipiscing elit. Ut purus elit, vestibulum ut, placerat ac, adipiscing vitae, felis. Curabitur dictum gravida mauris. Nam arcu libero, nonummy eget, consectetuer id, vulputate a, magna. Donec

**Fig. 2.9**  Example serial letter heading generated using Python

ing 2.39 - note that if your text contains commas, a different delimiter will have to be chosen (and passed to `pd.read_csv()`). Then, we iterate over the rows in this file, splitting up each item in a row (so passing the name and address to a variable with `name = row[0]` where row[0] indicates the 0-th item of the row [python is 0-indexed]), and so on. We can then use these new variables, by loading our LaTeX template (with `open(...)`), and using the function `replace`, to substitute our key with the desired value from our table, and so defining the commands that we set earlier. We then close the template, and save the edited document to a new individual file, making use of the string method `.split()`, to select the given and family name to be included in the filename.

```python
import pandas as pd

df = pd.read_csv('addresses.csv')
for index, row in df.iterrows():
    name = row[0]
    address = row[1]
    template = open("letter.tex", "rt")
    data = template.read()
    data = data.replace('#letterName', name)
    data = data.replace('#letterAddress', address)
    template.close()
    name = name.split()
    fileName = str("letter_"+str(name[1])+str(name[2])+".tex")
    letterFile = open(fileName, "wt")
    letterFile.write(data)
    letterFile.close()
```

**Listing 2.38**  Python code to generate serial letters

```
name,address
Ms. Jane Doe,10 Main Avenue
Mr. John Smith,15 Minor
Street
```

**Listing 2.39**  Basic CSV file used in the serial letter

This results in a letter header as shown in Fig. 2.9.

A further advantage of this approach is that one can make use of the many Python libraries, particularly those that integrate nicely with LaTeX. For example, using the tabulate Python library, one can generate tables that can be directly embedded into LaTeX. Using a similar approach as in the serial letters, the following Listing 2.40 shows a simplified approach to a report card, again generated with the help of a .csv file containing the grades for two students, shown in Listing 2.41.

```
1   \documentclass{article}
2   \usepackage{graphicx}
3
4   \newcommand{\student}{#studentName}
5
6   \begin{document}
7   \pagenumbering{gobble}
8
9   \large \begin{flushright}{\today}\end{flushright}
10  \noindent \Large{\textit\textbf{Report Card: }\student}\\
11
12  #table
13
14  \end{document}
```
**Listing 2.40**  Template for generating a basic report card

```
1   Subject,Anna Smith,James Redman
2   Analysis I,Very Good,High Distinction
3   Linear Algebra I,Very Good,Pass
4   Databases I,Pass,Distinction
5   Introduction to OOP,Very Good,Pass
```
**Listing 2.41**  Example CSV file used in the report card

In the LaTeX code there is no major change from the previous example, except that we do not define the a command for the table containing the example grades, as this is unlikely to be re-used multiple times in the document; hence, we simply use #table as a placeholder for the table of grades. In the Python code shown in Listing 2.42, some differences can be seen: we first load the table using pd.read_table() and store it in the variable df, and then iterate over the columns in this case (not over the rows as in the previous example), and select the entire column with df[student] (and convert back to a pandas dataframe, for tabulate to work properly). Using tabulate, we then convert this to table to LaTeX code. We then substitute as before, and then save the report card again with the student name.

```
1   import pandas as pd
2   import numpy as np
3   from tabulate import tabulate
4   df = pd.read_table('grades1.csv',sep=',',index_col=0, header='infer')
5   for student in df:
6       fin = open("reportCard.tex", "rt")
7       data = fin.read()
8       data = data.replace('#studentName', student)
9
10      grades = df[student].to_frame()
11      table = tabulate(grades, tablefmt = "latex")
12      data = data.replace('#table',table)
13      fin.close()
14
15      name=student.split()
16      fileName = str("reportCard_"+str(name[0])+str(name[1])+".tex")
17      fin = open(fileName, "wt")
18      fin.write(data)
19      fin.close()
```
**Listing 2.42**  Python code to generate LaTeX table code and generate report cards

**Fig. 2.10** Example report
card generated using Python

August 22,

Report Card: Anna Smith

| Analysis I | Very Good |
|---|---|
| Linear Algebra I | Very Good |
| Databases I | Pass |
| Introduction to OOP | Very Good |

This results in a sample report card as shown in Fig. 2.10.

If executing the Python scripts from the same folder as the LaTeX template, the generated .tex files will also be saved there. Once the .tex files have been generated, these can be compiled using the terminal, using e.g. pdflatex reportCard_*.tex, as this will compile all .tex files commencing with that filename. Note that these scripts have been written for Python 3.X, and require the installation of the pandas, numpy, and tabulate libraries to work properly.

## 2.7 Cross-Referencing and (PDF) Hyperlinks

Any section, figure, table, or equation, i.e. any numbered item, can be referenced to in LaTeX by giving the section a \label{}. You can then refer to these in the running text by using the command \ref{}, which will print the number of the item you are referencing. It is best common to give labels that are sufficiently descriptive, such as \label{table:geneTargets1}, which can then be referenced with \ref{table:geneTargets1}.

```
1   \section{Numbers} \label{sect:numbers}
2   \begin{table}[!ht]
3       \centering
4       \begin{tabularx}{\textwidth}{c|c}
5       1 & A\\
6       2 & B
7       \end{tabularx}
8       \caption{Number List}
9       \label{tab:numbers1}
10  \end{table}
11
12  As you can see in Table~\exref{tab:numbers1}...
13  [...]
14  The Table on page~\pageref{tab:numbers1}...
```

**Listing 2.43** Cross-referencing a table

It is important to place the \label{} command at the adequate position in order to reference the correct item. For \section{} and the like, the \label{} command must be placed right after that command. For figures and tables, the \label{} command must come after the \caption{} command. For equations, the \label{} command must come after the \begin{equation} command.

Additionally, one can not only print the number of a given item, but also what page that item is on, by using the command \pageref{label}.

It is common practice to use \ref{} in conjunction with a tilde (~), as with Figure~\ref{fig:results1}. The tilde (generally in LaTeX), prints a space that will not be broken by a new line. Hence, this prevents the words 'Figure', or 'Table', to be separated from the corresponding number.

With pdfTeX, some of the features of PDF documents can be used, such as hyperlinks - particularly useful for producing clickable table of contents, setting document information, and having clickable hyperlinks for URLs. There are two main packages that are used for this - hyperref and url.

The hyperref package allows for more advanced cross-referencing than just using \ref{}. For instance, one can link to any word in the document, by using the commands \hypertarget{key}{text} and \hyperlink{key}{text}. An example is shown in Listing 2.44.

```
1  This is the sentence I want to \hypertarget{keySentence}{target}.
2  [...]
3  This is the sentence where I want the reader to be sent to the above
4  \hyperlink{keySentence}{sentence}.
```
**Listing 2.44**  Usage of the hypertarget for cross-referencing

While in PDF, the usage of \label{} and \ref{} generates hyperlinks, allowing one to click on the number of the item being referenced and document jumping to that position, the words "Figure" or "Table", are not part of the hyperlink. To achieve this, one can simply use the command \hyperref[ref]{``Full text to be linked''}.

PDF documents can specify authors, titles, and keywords, which can be set when loading the hyperref package, as shown in Listing 2.45.

```
1  \usepackage[pdftex,
2    hidelinks,
3    pdfauthor={Author Name},
4    pdftitle={Title of Document},
5    pdfkeywords={Keywords}]{hyperref}
```
**Listing 2.45**  Setting PDF document information

To attach hyperlinks to URLs in LaTeX, the url package can be used. It is automatically loaded with hyperref, but can also be included in its own with \usepackage{url}. The package xurl additionally automatically allows for line breaks when including longer URLs. URLs are simply included by using \url{www.someurl.com}.

## 2.8  Additional Tools

While LaTeX facilitates the process of generating well typeset documents considerably, there are some tasks that at first may appear more complicated than with commonly used word processing software. A number of tools exist to help make these tasks easier.

### 2.8.1   Tables

A simple example here is making tables. In this chapter we have covered a number of packages that allow to precisely specify the formatting of tables. Nevertheless, sometimes it is only necessary to make a simple table, or to copy over tables from spreadsheets. For this, the website https://tablesgenerator.com/latex_tables is very useful, as it uses a graphical interface as with other spreadsheet programmes, and permits one to generate the corresponding LATEX code. Further, it allows one to convert data from .csv files (and thus can be used to import spreadsheet contents), allows a variety of edits to be made, as well as settings for LaTeX (centering, multi-page tables, width). This can also be used to generate a working table first, to then further edit, with packages such as booktabs for more detailed modifications. A similar tool can also be found at https://www.latex-tables.com/

An alternative may be to export spreadsheets, particularly those generated in MS Excel, to a latex compatible format directly. This is possible with the `excel2latex` package, available at https://ctan.org/pkg/excel2latex. An online version of such a tool is also available at http://excel2latex.com/.

### 2.8.2   Symbols

Sometimes it may be necessary to use a particular symbols in a document, for which a specific LATEX command may exist. To look this up, an online tool can be used, found at http://detexify.kirelabs.org/classify.html. For mathematical signs and symbols, a list can be consulted at https://en.wikipedia.org/wiki/List_of_mathematical_symbols_by_subject. A comprehensive collection of LATEX commands for symbols can be also be found at http://tug.ctan.org/info/symbols/comprehensive/symbols-a4.pdf.

### 2.8.3   Sketching

For specific purposes, it might be simplest to export drawings or sketches into a LATEX compatible format. Of course, any graphics can be simply included as a figure into LATEX; the advantage of having LATEX produce the graphics is that any fonts included can be managed from within LATEX and that the size and positioning can be adjusted more finely. Alternatively, graphics software like Inkscape is capable of separately exporting images from fonts, allowing any text to be rendered by LATEX.

A simple GUI tool that allows the drawing of basic graphs and diagrams is Tikzit. It generates Tikz/PGF output code that can be directly included in LATEX text, and can be found at https://tikzit.github.io/.

For particularly mathematical drawings, sketches, geometry, and other diagrams, https://www.mathcha.io/ offers an easy to use graphical environment that exports easily to LaTeX code. For chemical structures, an online application to draw and export chemical structures can be found at https://py-chemist.com/mol_2_chemfig/ home.

To draw more complex graphics, Inkscape is an open source vector graphics editor that allows exporting to PDF, with an option to exclude all text from the PDF figure, and include based on generated LaTeX code. This has the advantage of allowing any text required in figures to share the same font as the main document. It can be downloaded at https://inkscape.org/.

# Chapter 3
# Floating Objects

**Abstract** This chapter introduces the reader to the graphical representation of numerical data based on the package PGFPLOTS. Furthermore, the automated generation of tables from external ASCII files is explained. At the end, possibilities for spreadsheet calculations and statistical analyses are presented.

## 3.1 Figure Generation with PGFPLOTS

### 3.1.1 Basics

The LaTeX package PGFPLOTS allows the plotting of functions and other representations of numerical data [12]. The package must be included in the document preamble as shown in Listing 3.1 below:

```
1   \documentclass[...]{....}
2   \usepackage{pgfplots}
3   ...
4   \begin{document}
5   ...
```

**Listing 3.1** Incorporation of the PGFPLOTS package

A typical environment to plot two analytical functions is generally as shown in Listing 3.2.

```
1   \begin{figure}
2       \begin{tikzpicture}
3           \begin{axis}[option_a_1,option_a_2,...]
4           \addplot[option_p_1,...] {function_1};
5           \addplot[option_p_1,...] {function_2};
6           \end{axis}
7       \end{tikzpicture}
8   \end{figure}
```

**Listing 3.2** The basic structure of a PGFPLOTS environment

The most common way to implement a figure in LaTeX is to begin with a floating figure environment (\begin{figure}...\end{figure}), which ensures that a figure

M. Öchsner and A. Öchsner, *Advanced LaTeX in Academia*,
https://doi.org/10.1007/978-3-030-88956-2_3

**Fig. 3.1** Graphical
representation of a parabola,
see Listing 3.3

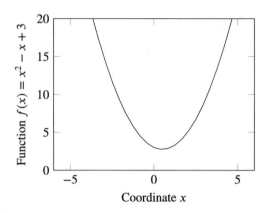

number and caption can be assigned to the figure. The actual environment for the
PGFPLOTS figure is enclosed in \begin{tikzpicture}...\end{tikzpicture}, which is
followed by a \begin{axis}...\end{axis} environment to represent data in a coordinate system. The command \addplot[...]{...} allows to define a function to be plotted.

The following example shows the graphical representation of a parabola ($y(x) =$
$x^2 - x + 3$) and simple formatting options (see the following Listing 3.3 and
Fig. 3.1).

```
1   \begin{figure}\centering
2       \begin{tikzpicture}
3           \begin{axis}[
4           width=0.55\textwidth, height=0.44\textwidth,
5           xlabel={Coordinate $x$}, ylabel={Function $f(x)
6           = x^2 - x +3$},xmin=-6, xmax=6,ymin=0, ymax=20,]
7           \addplot[mark=none]  {x^2 - x + 3};
8           \end{axis}
9       \end{tikzpicture}
10  \end{figure}
```

**Listing 3.3**  Simple PGFPLOTS example of a parabola

The axes options width=... and height=... allow to specify the outer dimensions of
the diagram while xlabel={...} and ylabel={...} allow to add axes labels. The plot
option mark=none indicates that no markers are used to represent the course of the
function.

Table 3.1 summarizes common commands for defining and formatting the axes
of a graph. These basic commands are already sufficient to create a simple graph. In
the case that the automated axes numbering style must be changed, Table 3.2 collects
several commands to adjust the display of the axes numbers.

Tables 3.3, 3.4 and 3.5 summarize different formatting options for the command
\addplot.

Let us consider in the following a data table with the file name data_table_1.txt,
which is located in a subdirectory ('tables') of the actual working directory. This
ASCII file contains four columns which represent a single $x$-axis and three different
$y$-axes.

**Table 3.1** Commands for defining and formatting of axes (option_a)

| Category | option_a | Comment |
|---|---|---|
| Dimensions | width=0.55\textwidth | Defines plot width in relation to text width |
| | width=7cm | Defines plot width in absolute length |
| | height=0.44\textwidth | Defines plot height in relation to text width |
| Axes Labels | xlabel={Coordinate $x$} | Axis label for horizontal axis |
| | ylabel={Function $y(x)$} | Axis label for vertical axis |
| | label style={font=\tiny} | Sets the axes font size to tiny |
| | y label style={at={(axis description cs:0.075,0.5)}} | Allows to adjust the position of the vertical axis label in $x$- and $y$-direction |
| Axes Ranges | xmin=-6, xmax=6 | Sets the range of the horizontal axis from $-6$ to 6 |
| | ymin=0, ymax=20 | Sets the range of the vertical axis from 0 to 20 |
| Axes Ticks | xtick={0,0.1,0.3,0.4,0.5} | Draws the horizontal axis tick only at coordinates 0, 0.1, 0.3, 0.4 and 0.5 |
| | xtick=\empty | Tick are suppressed for horizontal axis |
| | xticklabels={0,$a$,$\frac{a}{b}$,0.4,0.5} | Assigns a label to each tick position. Tick positions are defined by xtick={...} |
| | tick label style={font=\tiny} | Sets the tick label font size to tiny |
| | x tick label style={...} | Allows number formatting for horizontal axis, see Table 3.2 |
| | scaled y ticks = false | Prevents using the $10^n$ notation for the $y$-axis (axis multiplier) |
| | scaled ticks = false | Prevents using the $10^n$ notation for both axes (axes multipliers) |

```
x_a    y_1    y_2    y_3
0      0.0    1.0    1.1
1      0.5    1.2    1.3
2      1.0    1.4    1.5
3      1.5    1.6    1.7
4      2.0    1.8    1.9
5      2.5    2.0    2.1
```

The following Listing 3.4 explains the plotting of three curves based on the external ASCII file (see Fig. 3.2). The LaTeX package **pgfplotstable** is required for this approach. The command **\pgfplotstableread** reads the specified external file and saves it under the variable **\datatable**. The selection of the different markers is explained in Table 3.7.

```
1  \begin{figure}\centering
2  \pgfplotstableread{tables/data_table_1.txt}\datatable
3      \begin{tikzpicture}
4      \begin{axis}[width=0.55\textwidth, height=0.44\textwidth,
5          xlabel={Coordinate $x$}, ylabel={Data sets $y_i$},
```

**Table 3.2** Commands for defining the number formatting (e.g. x tick label style={/pgf/number format/.cd,option_n}). Adapted from [13]

| option_n | Comment |
| --- | --- |
| fixed | Round the number to a fixed number of digits after the period, discarding any trailing zeros |
| sci | To display numbers in scientific format, i.e. sign, mantissa and exponent to base 10 |
| fixed zerofill | Enables zero filling for any number drawn in fixed point format. Use fixed zerofill=false to disable |
| sci zerofill | Enables zero filling for any number drawn in scientific format. Use sci zerofill=false to disable |
| zerofill | Sets both fixed zerofill and sci zerofill at once. Use zerofill=false to disable |
| precision=2 | Sets the desired rounding precision for any display operation (here: 2). For scientific format, this affects the mantissa. Use in conjunction with other commands, for example: fixed, fixed zerofill, precision=2 |
| sci precision=2 | Sets the desired rounding precision only for sci styles (here: 2). Use in conjunction with other commands, for example: sci, sci zerofill, precision=2 |
| frac | Displays numbers as fractionals |
| set decimal separator={ } | Defines the decimal separator for any fixed point number (including the mantissa in sci format). For example for a comma: set decimal separator ={,\!} |
| set thousands separator={ } | Defines the thousands separator for any fixed point number (including the mantissa in sci format). For example for a period: set thousands separator={.}. For example no separator: set thousands separator={ } |
| use period | A predefined style which used periods '.' as decimal separators and commas ',' as thousands separators |
| use comma | A predefined style which installs commas ',' as decimal separators and periods '.' as thousands separators |
| showpos=true | Displays the plus signs for non-negative numbers. Default is false |
| sci generic={mantissa sep=\times,exponent= {10^{#1}}} | Uses in scientific notation \times as the mantissa separator instead of the default \cdot. #1 is the actual exponent. One may replace it by any integer, e.g. 3. This would mean that all numbers would have the factor $10^3$ |
| tick scale binop=\times | Uses \times as the axis multiplier instead of the default \cdot |

**Table 3.3**  Different plot options in PGFPLOTS (option_p)

| option_n | Comment |
|---|---|
| line width=5pt | Defines the line thickness of the plotted function. The default value is 0.4 pt |
| domain=-10:10 | Defines the $x$-range for which the function is plotted |
| samples=100 | Defines the number of points for which the function is evaluated |
| color=red | Defines the line color of the plotted function (see Table ??) |

**Table 3.4**  Different line styles in PGFPLOTS (option_p). The same commands are available in a regular TikZ picture

| Name | Display | Name | Display |
|---|---|---|---|
| solid | ——— | dotted | ············ |
| densely dotted | ···················· | loosely dotted | · · · · · · · |
| dashed | – – – – – | densely dashed | – – – – – – |
| loosely dashed | – – – – | dashdotted | –· –· –· –· |
| densely dashdotted | –·–·–·–·– | loosely dashdotted | – · – · – |
| dashdotdotted | – ·· – ·· – ·· | densely dashdotdotted | –··–··–··– |
| loosely dashdotdotted | – · · – · · | | |

**Table 3.5**  Options to define line thickness in PGFPLOTS and TikZ (option_p)

| option_n | Comment |
|---|---|
| ultra thin | Corresponds to 0.1 pt |
| very thin | Corresponds to 0.2 pt |
| thin | Corresponds to 0.4 pt (default value) |
| semi thick | Corresponds to 0.6 pt |
| thick | Corresponds to 0.8 pt |
| very thick | Corresponds to 1.2 pt |
| ultra thick | Corresponds to 1.6 pt |
| line width = 0.5mm | Corresponds to 0.5 mm |
| line width = 0.5 | Corresponds to 0.5 pt |

**Fig. 3.2** Plotting of data
points from an external file,
see Listing 3.4

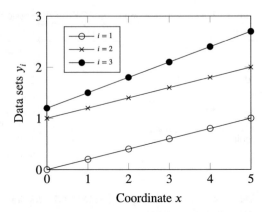

Coordinate $x$

```
6                        xmin=0, xmax=5,ymin=0, ymax=3,xtick={0,1,2,3,4,5},
7                        legend style={at={(axis cs:1.8,2.8)},font=\tiny}]
8           \addplot[mark=o,color=black] table[x = x_a, y = y_1] from \datatable;
9           \addplot[mark=x,color=black] table[x = x_a, y = y_2] from \datatable;
10          \addplot[mark=*,color=black] table[x = x_a, y = y_3] from \datatable;
11          \legend{$i=1$,$i=2$,$i=3$}
12        \end{axis}
13      \end{tikzpicture}
14  \end{figure}
```

**Listing 3.4**  Plotting of data points from an external file

An alternative way to read data from an external file is the direct incorporation in the
\addplot command as shown in Listing 3.5 (lines 7–9). This approach works even
without the LaTeX package pgfplotstable.

```
1   \begin{figure}\centering
2   \begin{tikzpicture}
3   \begin{axis}[width=0.55\textwidth, height=0.44\textwidth,
4               xlabel={Coordinate $x$}, ylabel={Data sets $y_i$},xmin=0,
5               xmax=5,ymin=0, ymax=3,xtick={0,1,2,3,4,5},
6               legend style={at={(axis cs:1.8,2.8)},font=\tiny}]
7   \addplot[mark=o,color=black] table[x = x_a, y = y_1] {tables/data_table_1.txt};
8   \addplot[mark=x,color=black] table[x = x_a, y = y_2] {tables/data_table_1.txt};
9   \addplot[mark=*,color=black] table[x = x_a, y = y_3] {tables/data_table_1.txt};
10  \legend{$i=1$,$i=2$,$i=3$}
11  \end{axis}
12  \end{tikzpicture}
13  \caption{Plotting of data points from external file}
14  \label{fig:plotting_of_data_from_external_file}
15  \end{figure}
```

**Listing 3.5**  Plotting of data points from an external file (alternative approach)

Let us have now a look at a larger data set. The set is saved in a data table with the
file name **data_table_pgf_1.txt**, which is located in a subdirectory ('tables') of the
actual working directory.

| x_1 | y_1 |
|---|---|
| 0 | 1 |
| 1 | 1.4 |
| 2 | 1.4 |
| 3 | 1.4 |
| 4 | 1.8 |

```
 5    2.3
 6    2.3
 7    2.5
 8    2.7
 9    2.9
10    3.0
11    3.3
12    3.5
13    3.9
14    4.0
15    4.5
16    4.7
17    4.8
18    5.0
19    5.2
20    5.3
```

The following explains how to reduce the plotted data or select only specific ranges of the entire data set. The procedure to plot the entire set of data points is given in Listing 3.6, see Fig. 3.3a.

```
1    ...
2    \begin{tikzpicture}
3        \begin{axis}[width=1.0\textwidth, height=0.9\textwidth,
4                     xlabel={Coordinate $x$}, ylabel={Data set $y$},xmin=0,
5                     xmax=20,ymin=0, ymax=6]
6            \addplot[mark=o,color=black] table [x=x_1,y =y_1] {tables/
7                                                     data_table_pgf_1.txt};
8        \end{axis}
9    \end{tikzpicture}
10   ...
```

**Listing 3.6** Plotting of data points from an external file (all data points, see Fig. 3.3a)

To reduce the plotted data, the following listings show the modifications of line 6 in Listing 3.6.

```
6    \addplot[mark=o,color=black,each nth point = 3] table [x=x_1,y=y_1] {tables/
7                                                     data_table_pgf_1.txt};
```

**Listing 3.7** Plotting of data points from an external file (each third point, see Fig. 3.3b)

```
6    \addplot[mark=o,color=black,skip coords between index = {0}{9}]
7        table [x=x_1,y=y_1] {figs/03_figs/data_table_pgf_1.txt};
```

**Listing 3.8** Plotting of data points from an external file (data rows from 0 to 9 omitted, see Fig. 3.3c)

```
6    \addplot[mark=o,color=black,skip coords between index = {0}{9},skip coords
7        between index = {15}{21}] table [x = x_1, y = y_1] {figs/03_figs/
8        data_table_pgf_1.txt};
```

**Listing 3.9** Plotting of data points from an external file (data rows from 0 to 9 and from 15 to 21 omitted, see Fig. 3.3d)

The legend as shown in Fig. 3.2 (see the corresponding Listing 3.4) can be easily customized. See Table 3.6 for some options. Some commands to customize the use of markers are given in Tables 3.7 and 3.8.

Experimental data analyses generally require statistical evaluation and the graphical representation requires, in general, the indication of error bars (Fig. 3.4 and the corresponding Listing 3.10). The use of error bars is indicated by error bars... as an option in the addplot command. The deviation itself is indicated by +-($\Delta x, \Delta y$) after each data point. The $\Delta x$ relates to the error in regards to the $x$-coordinate whereas

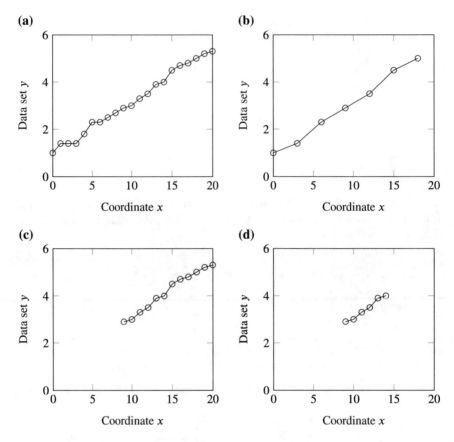

**Fig. 3.3** Selective reading of table data: **a** all data points, **b** each third point, **c** data rows from 0 to 9 omitted, and **d** data rows from 0 to 9 and from 15 to 21 omitted. See Listings 3.6–3.9

$\Delta y$ relates to the error with respect to the $y$-coordinate. If the deviation is not the same in the positive and negative direction, the commands +=() and -=() allow to distinguish the direction.

```
1   \begin{figure}\centering
2   \begin{tikzpicture}
3       \begin{axis}[width=0.55\textwidth, height=0.44\textwidth,
4           xlabel={Coordinate $x$}, ylabel={Data set $y$},
5           xmin=-0.2, xmax=5.2,ymin=0, ymax=3,xtick={0,1,2,3,4,5}]
6           \addplot[mark=x,color=black, error bars/.cd,x dir=both,x explicit,
7               y dir=both,y explicit] plot[] coordinates{(0,1)+-(0,0.2)
8               (1,1.2)+=(0.1,0.2)-=(0.2,0.4) (2,1.4)+-(0.5,0.4) (3,1.6)
9               +=(0.2,0.2)-=(0.1,0.1) (4,1.8)+-(0.5,0.3) (5,2.0)+-(0.1,0.2)};
10      \end{axis}
11  \end{tikzpicture}
12  \caption{Plotting of a data set with $x$- and $y$-error bars}
13  \label{fig:coord_error_bars}
14  \end{figure}
```

**Listing 3.10**  Plotting of a data set with $x$- and $y$-error bars

**Table 3.6** Different options to adjust the legend style in PGFPLOTS (option_a)

| option_a | Comment |
| --- | --- |
| legend entries = {$d=2$, $d=3$, $d=4$} | Defines the legend entries for 3 curves (3 times \addplot...) |
| legend style={draw=none} | Without a box around the legend |
| legend pos = south east | Defines the legend position in side the graph (also possible: north, west, outer) |
| legend style={at = {(axis cs:1.8,2.8)}} | Defines the legend position with reference to the axis coordinate system |
| legend style = {overlay, at = {(-0.5,0.5)}, anchor = center} | Defines the legend position without affecting the bounding box |
| legend style = {cells = {anchor = west}} | Defines the text alignment: west = right, east = left, center |
| legend columns =-1 | Draws all legend entries horizontally. 2: max. two adjacent legend entries. 1: Draws all legend entries vertically below each other (default) |
| legend style = {font = \small} | To adjust the font size |

**Fig. 3.4** Plotting of a data
set with $x$- and $y$-error bars,
see Listing 3.10

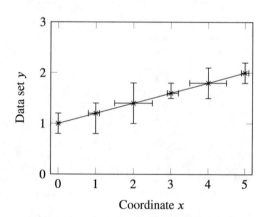

Coordinate $x$

Some graphical representations of data requires that axis with discontinuities are drawn. The built-in function **axis y discontinuity** (or corresponding for the $x$-axis) allows to indicate that the coordinate origin is not included, see Fig. 3.5 and Listing 3.11.

```
\begin{figure}
\centering
    \begin{subfigure}[t]{0.48\textwidth}
    \centering
    \captionsetup{labelfont=bf}
    \caption[]{\hspace*{5cm}}
    \begin{tikzpicture}
    \begin{axis}[width=1.00\textwidth, height=0.9\textwidth,
                xlabel={Coordinate $x$}, ylabel={Function $y(x)$},xmin=-6,
                xmax=6,ymin=0, ymax=30,  axis y discontinuity=crunch]
    \addplot+[mark=none,domain =-5:5,color=black] {x^2 +8};
```

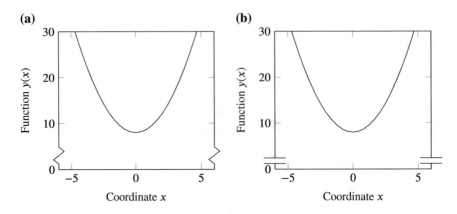

**Fig. 3.5**  Axis discontinuity: **a** crunch and **b** parallel style, see Listing 3.11

```
12      \end{axis}
13      \end{tikzpicture}
14      \end{subfigure}
15      \begin{subfigure}[t]{0.48\textwidth}
16      \centering
17      \captionsetup{labelfont=bf}
18      \caption[]{\hspace*{5cm}}
19      \begin{tikzpicture}
20      \begin{axis}[width=1.00\textwidth, height=0.9\textwidth,
21                 xlabel={Coordinate $x$}, ylabel={Function $y(x)$},xmin=-6,
22                 xmax=6,ymin=0, ymax=30,  axis y discontinuity=parallel]
23      \addplot+[mark=none,domain =-5:5,color=black] {x^2 +8};
24      \end{axis}
25      \end{tikzpicture}
26      \end{subfigure}
27  \caption{Axis discontinuity: \textbf{(a)} crunch and \textbf{(b)}
28              parallel style}
29  \label{fig:pgf_axis_disonti}
30  \end{figure}
```

**Listing 3.11**  Axis discontinuity: (a) crunch and (b) parallel style

A more advanced problem is a discontinuity at an arbitrary location of the $y$-axis. This can be achieved based on two group plots, see Fig. 3.6 and Listing 3.12. This example uses the groupplots library of the PGFPLOTS package. The idea is to define a common $x$-axis and then two sets of $y$-axes with different ranges. These two sets are defined with the command \nextgroupplot.

```
1   \documentclass{article}
2   \usepackage{pgfplots}
3   \usetikzlibrary{pgfplots.groupplots}
4   ...
5   \begin{document}
6   ...
7   \begin{figure}
8   \centering
9       \begin{tikzpicture}
10      \begin{groupplot}[group style={group name=disc plots,group size=1 by 2,
11                  xticklabels at=edge bottom,vertical sep=0pt},
12                  width=0.55\textwidth, height=0.44\textwidth,xmin=-6,
13                  xmax=6, every non boxed x axis/.style={}]
```

**Fig. 3.6** Arbitrary axis
discontinuity, see
Listing 3.12

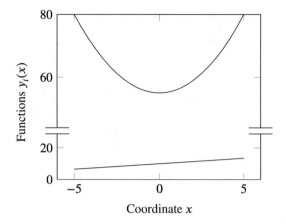

```
14    \nextgroupplot[ymin=40,ymax=80,ytick={60,80},axis x line=top,
15                   axis y discontinuity=parallel,height=4.5cm,ylabel=
16                   {Functions $y_i(x)$},y label style={at={(axis description
17                   cs:0.05,0.3)}}]
18    \addplot[domain =-5:5,mark=none,color=black] {0.7*x+10};
19    \addplot[domain =-5:5,mark=none,color=black] {x^2+55};
20    \nextgroupplot[ymin=0,ymax=25,ytick={0,20},axis x line=bottom,height=2.5cm,
21                   xlabel={Coordinate $x$}]
22    \addplot[domain =-5:5,mark=none,color=black] {0.7*x+10};
23    \addplot[domain =-5:5,mark=none,color=black] {x^2+55};
24    \end{groupplot}
25    \end{tikzpicture}
26    \caption{Arbitrary axis discontinuity}
27    \label{fig:pgf_axis_disonti_arb}
28    \end{figure}
```

**Listing 3.12** Arbitrary axis discontinuity

Some representation of data requires that a second $y$-axis is displayed, see Fig. 3.7 and Listing 3.13. The second axis requires using two layers, which is indicated by the pgfplot setting 'set layers'. Then, two \axis environments are used to define a set with a left (indicated by axis y line*=left) and a set with a right $y$-axis (indicated by axis y line*=right). The command ylabel near ticks ensures that the axis label of the second axis is plotted near the corresponding axis. If only one function is plotted, it is sufficient to define the $x$-axis in one set. If two functions are plotted, the same $x$-axes should be defined in both sets.

```
1     \begin{figure}
2     \centering
3         \begin{subfigure}[t]{0.48\textwidth}
4         \centering
5         \captionsetup{labelfont=bf}
6         \caption[]{\hspace*{5cm}}
7         \begin{tikzpicture}
8         \pgfplotsset{ set layers}
9         \begin{axis}[scale only axis,width=1.0\textwidth, height=0.9\textwidth,
10                    axis y line*=left,xlabel={Coordinate $x$}, ylabel=
11                    {Function $y(x)$},xmin=-6, xmax=6,ymin=0, ymax=30]
12        \addplot[mark=none,domain =-5:5,color=black] {x^2 + 5};
13        \end{axis}
14        \begin{axis}[scale only axis,width=1.0\textwidth, height=0.9\textwidth,
```

**Fig. 3.7**  Second *y*-axis: **a** single function and **b** related to a second function, see Listing 3.13

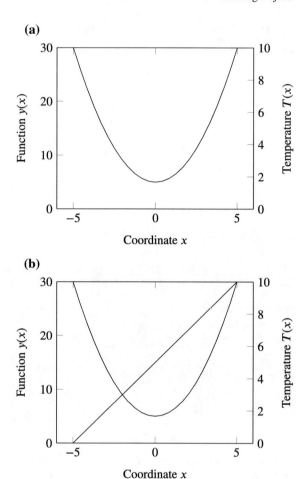

(a)

(b)

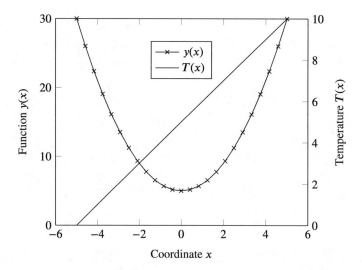

**Fig. 3.8** Representation of two functions with two $y$-axes and joined legend, see Listing 3.14

```
35    \end{axis}
36    \end{tikzpicture}
37  \end{subfigure}
38  \caption{Second $y$-axis: \textbf{(a)} single function and \textbf{(b)} related
39          to a second function}
40  \label{fig:pgf_sec_y_axis}
41  \end{figure}
```

**Listing 3.13** Second $y$-axis: (a) single function and (b) related to a second function

In case that a combined legend (see Fig. 3.8) is required, the approach in Listing 3.14 can be used: The first plot is given a label (\label{plot_temp}), which is used in the second axis environment to define the combined legend. Important is that the command \addlegendimage is introduced before the second plot command (\addplot).

```
1   \begin{figure}
2   \centering
3     \begin{tikzpicture}
4     \pgfplotsset{set layers}
5     \begin{axis}[scale only axis, width=0.55\textwidth,
6                  height=0.44\textwidth, axis y line*=left,
7                  xlabel={Coordinate $x$}, ylabel={Function $y(x)$},
8                  xmin=-6, xmax=6, ymin=0, ymax=30]
9     \addplot[mark=x,domain =-5:5,color=black] {x^2 + 5};
10    \label{plot_temp}
11    \end{axis}
12    \begin{axis}[scale only axis, width=0.55\textwidth,
13                 height=0.44\textwidth, xmin=-6, xmax=6,ymin=0,
14                 ymax=10, axis y line*=right, axis x line=none,
15                 ylabel={Temperature $T(x)$}, ylabel near ticks,
16                 legend style={at={(rel axis cs:0.5,0.8)},anchor=center}]
17    \addlegendimage/pgfplots/refstyle=plot_temp\addlegendentry{$y(x)$}
18    \addplot[mark=none,domain =-5:5,color=black] {x + 5};
19    \addlegendentry{$T(x)$}
20    \end{axis}
21    \end{tikzpicture}
22  \caption{Representation of two functions with two $y$-axes and joined legend}
```

**Table 3.7** A selection of markers in PGFPLOTS and TikZ (option_p). The additional option only marks avoids joining lines between markers

| mark= | Marker | Comment | mark= | Marker | Comment |
|---|---|---|---|---|---|
| + | + | | - | — | |
| x | × | | | | |
| o | ○ | unfilled | * | ● | filled |
| star | ⋆ | | asterisk | ∗ | |
| triangle | △ | unfilled | square | □ | unfilled |
| triangle* | ▲ | filled | square* | ■ | filled |
| diamond | ◇ | unfilled | pentagon | ⬠ | unfilled |
| diamond* | ◆ | filled | pentagon* | ⬟ | filled |
| none | | | | | |

**Table 3.8** Commands for defining the layout of markers (option_p)

| Category | option_p | Comment |
|---|---|---|
| Size | mark size=0.6pt | |
| | mark options={scale=2} | |
| Color | mark options={color=blue} | Unfilled markers: changes only the contour. Filled markers: changes contour and filling |
| | mark options={fill=green} | Filled markers: changes only the filling ('fill') |
| | mark options={color=blue, fill=green} | Filled markers: changes the contour ('color') and the filling ('fill') |
| Line Style | mark options={solid} | Sets the marker line style to 'solid' in case that the graph is plotted, for example, as dotted |
| Others | mark options ={rotate=90} | Rotates the marker, for example, by 90° |
| | mark=text,text mark=A | Assigns, for example, the alphabetic character 'A' as marker |
| | text mark as node=true,text mark style={font=\tiny} | Allows to define the style of the text mark from the previous command |

```
23    \label{fig:pgf_sec_y_axis_legend}
24    \end{figure}
```

**Listing 3.14** Representation of two functions with two $y$-axes and joined legend

TikZ allows to add additional objects to a PGFPLOTS graph as shown in Table 3.9. In case that a coordinate must be defined, the specification (axis cs:x,y) allows to refer to the displayed axes of the plot. Alternatively, a normalized coordinate system can be used by (rel axis cs:x,y) in which the lower left corner is (0,0) and upper right corner is (1,1). It holds $0 \leq x \leq 1$ and $0 \leq y \leq 1$ for the relative coordinate system.

**Table 3.9** TikZ commands for drawing simple objects in PGFPLOTS

| Category | Command | Comment |
|---|---|---|
| Shapes | \draw (axis cs:0, 0) rectangle (axis cs:1, 0.5) | Draws an empty rectangle with the lower left corner (0,0) and the upper right corner (1, 0.5) with reference to the coordinate system of the plot (as displayed) |
| | \draw[option_d] (axis cs:0, 0) rectangle (axis cs:1, 0.5) | Allows to specify draw options |
| | \filldraw[fill=red, draw = blue] (axis cs:2 ,0) rectangle (axis cs:3 ,1) | Draws a filled rectangle with the filling color 'fill=red' and border line color 'draw = blue'. Use this, for example, to create partial backgrounds (specify the same color for filling and border) |
| | \draw[color=red] (axis cs:0, 0) circle (.1) | Draws a circle with center (0,0) and radius 0.1 |
| | \draw[option_d] (axis cs:0,0) - - (axis cs:0,2) | Draws a line between the coordinates (0,0) and (0,2) |
| | \draw[option_d] (axis cs:0,0) - - (axis cs:0,2) - - (axis cs:1,4) | Draws a line from the coordinate (0,0) to (0,2) and then to (1,4) |
| | \draw[option_d] (axis cs:0,0) - - +(axis direction cs:1,2) | Use of relative coordinates: draws a line from the coordinate (0,0) to the point (0+1,0+2) |
| Nodes etc. | \node[option_n] at (axis cs:x,y) {text} | Centers the word 'text' at the coordinate $(x, y)$. See Table 3.10 for option_n |
| | \node[option_n] at (axis cs:x,y) {\begin{small} $\frac{3q_0 L}{28}$ \end{small}} | Centers the fraction $\frac{3q_0 L}{28}$ in small font size at the coordinate $(x, y)$ |
| | \node [] at (axis cs:x,y) \includegraphics[scale=0.5] {fig_name}; | To place an external figure called fig_name in a PGFPLOT |
| | \node (A) at (axis cs:0,1) {...}; | Assigns to the node (0,1) the name (A) |
| | \coordinate (A) at (axis cs:0,1); | Assigns to the coordinate (0,1) the name (A). (A) can be used instead of the coordinate values |

**Table 3.10** Different options for nodes (option_n)

| option_n | Comment |
|---|---|
| scale=0.4 | Scales the size of the displayed node text to 40% of the original size |
| rotate=90 | Rotates the displayed node text by 90° |
| draw | Draws a rectangle around the displayed node text |
| inner sep=3mm | Defines the space around the displayed node text (to the surrounding 'object') |
| rectangle | Surrounds the the displayed node text by a rectangle. Use rounded corners to round the corners |
| circle | Surrounds the the displayed node text by a circle |
| color=red | Sets the color for the node to red, see Table A.2 |
| fill=blue | Fills the surrounding 'object' with the color blue, see Table A.2 |
| semitransparent | Semitransparent filling color |
| below | Places the node text below the given node coordinates. Alternatives are 'above', 'left', 'right' or even combinations such as 'below left' (i.e., the upper right corner of the surrounding rectangle would be at the node coordinate) |
| below=5mm | Allows to define an offset. Also works also when combining options, e.g. below left=5mm |
| text width=5cm | Manually defines the width of the node. Use align=center to center the text |

The option_d in conjunction with the draw command (see Table 3.9) may be used to specify the line style (see Table 3.4), the line width via the option line width=... (the default value is: line width=0.4pt), or the color via the option color=... (see Table A.2).

Let us combine is the following several plotting functions. Figure 3.9a shows the combination of a line (see Table 3.11) with a node (see Table 3.9). The option pos=.5 places the node in the middle of the line object while the option sloped rotates the node according to the slope of the line (see line 13 of the Listing 3.15).

```
1   \begin{figure}
2   \centering
3           \begin{subfigure}[t]{0.48\textwidth}
4           \centering
5           \captionsetup{labelfont=bf}
6           \caption[]{\hspace*{5cm}}
7                   \begin{tikzpicture}
8                   \begin{axis}[width=1.00\textwidth,height=0.90\textwidth,
9                   xlabel={Coordinate $x$},ylabel={Function $y(x)$},
10                  xmin=0, xmax=6.0,ymin=0.0, ymax=20.0]
11                  \coordinate (A) at (axis cs:1.0,5);
12                  \coordinate (B) at (axis cs:4.0,15);
13                  \draw[|{latex}-{latex}|](A)--(B)node[pos=.5,sloped,above]{$L$};
14                  \end{axis}
15                  \end{tikzpicture}
16          \end{subfigure}
17          \begin{subfigure}[t]{0.48\textwidth}
18          \centering
19          \captionsetup{labelfont=bf}
```

**Table 3.11** TikZ commands for drawing lines in PGFPLOTS

| Command | Display |
|---|---|
| \draw (axis cs:0, 0) - - (axis cs:0, 2) | |
| \draw[->] (axis cs:0, 0) - - (axis cs:0, 2) | |
| \draw[<->] (axis cs:0, 0) - - (axis cs:0, 2) | |
| \draw[|<->|] (axis cs:0, 0) - - (axis cs:0, 2) | |
| \draw[-»] (axis cs:0, 0) - - (axis cs:0, 2) | |
| \draw[latex-latex] (axis cs:0, 0) - - (axis cs:0, 2) | |
| \draw[|{latex}-{latex}|] (axis cs:0, 0) - - (axis cs:0, 2) | |
| \draw[stealth-stealth] (axis cs:0, 0) - - (axis cs:0, 2) \usetikzlibrary{arrows} | |
| \draw[)-o] (axis cs:0, 0) - - (axis cs:0, 2) | |
| \draw[*-angle 45] (axis cs:0, 0) - - (axis cs:0, 2) | |
| \draw[angle 60-angle 60] (axis cs:0, 0) - - (axis cs:0, 2) | |
| \draw[angle 90-angle 90] (axis cs:0, 0) - - (axis cs:0, 2) | |
| \draw[triangle 45-triangle 45] (axis cs:0, 0) - - (axis cs:0, 2) | |
| \draw[triangle 60-triangle 60] (axis cs:0, 0) - - (axis cs:0, 2) | |
| \draw[triangle 90-triangle 90] (axis cs:0, 0) - - (axis cs:0, 2) | |
| \draw[triangle 90 reversed-triangle 90 reversed] (axis cs:0, 0) - - (axis cs:0, 2) | |
| \draw[latex'-stealth'] (axis cs:0, 0) - - (axis cs:0, 2) | |
| \draw[open triangle 60-open triangle 90 reversed] (axis cs:0, 0) - - (axis cs:0, 2) | |

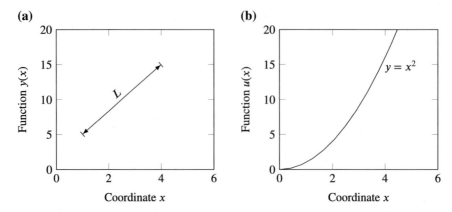

**Fig. 3.9** Combining features: **a** adding a node to a drawn dimension line **b** adding a node to a plotted curve, see Listing 3.15

```
20        \caption[]{\hspace*{5cm}}
21                \begin{tikzpicture}
22                \begin{axis}[width=1.00\textwidth,height=0.90\textwidth,
23                xlabel={Coordinate $x$},ylabel={Function $u(x)$},
24                xmin=0, xmax=6.0,ymin=0.0, ymax=20.0]
25                \addplot[] {x^2} node[pos=0.8,right] {$y=x^2$};
26                \end{axis}
27                \end{tikzpicture}
28        \end{subfigure}
29    \caption{Combining features: \textbf{(a)} adding a node to a drawn dimension
30    line \textbf{(b)} adding a node to a plotted curve}
31    \end{figure}
```

**Listing 3.15**  Combination of plotting functions

Figure 3.9b shows the combination of a plotted function with a node (see line 25 of the listing). The position of the node is defined at 80% of the length of the function and on the right-hand side (pos=0.8,right). Figure 3.9 shows in addition the combination of two figures in a single figure environment. This is done based on the subfigure environment and requires that the following library (groupplots) and package (subcaption) are included at the beginning of the document, see Listing 3.16.

```
1    \documentclass ....
2    ...
3    \usepackage{pgfplots}
4    \usetikzlibrary{pgfplots.groupplots}
5    \usepackage[]{subcaption}
6    ...
7    \begin{document}
8    ...
```

**Listing 3.16**  Setting for several figures as a group plot

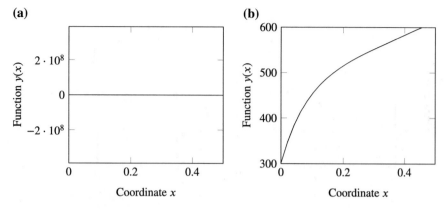

**Fig. 3.10**  Plotting of a comprehensive function: **a** without specification of the domain and **b** under specification of the domain, see Listings 3.17 and 3.18

### 3.1.2  Comments on Plotting Functions

Let us highlight in the following some issues which may occur during the plotting of functions. Equation (3.1) shows a polynomial of order six.

$$y(x) = 301.0 + 2455.4x - 13304.0x^2 + 49557.0x^3 - 115840.0x^4$$
$$+ 154820.0x^5 - 88689.9x^6 \tag{3.1}$$

Plotting the function without specification of further options as indicated in Listing 3.17 results in a quite unsatisfactory result as shown in Fig. 3.10a.

```
1   ...
2   \addplot[mark=none] {301.0+2455.4*x−13304.0*x^2+49557.0*x^3−115840.0*x^4
3                        +154820.0*x^5−88689.9*x^6};
4   ...
```

**Listing 3.17**  Plotting of a comprehensive function without specification of the domain

An improvement is obtained by specifying the plot domain as shown in Listing 3.18, see Fig. 3.10b.

```
1   ...
2   \addplot[domain=0:0.5,mark=none] {301.0+2455.4*x−13304.0*x^2+49557.0*x^3
3                                     −115840.0*x^4+154820.0*x^5−88689.9*x^6};
4   ...
```

**Listing 3.18**  Plotting of a comprehensive function under specification of the domain

The graphical representation of a function with a pole (see Eq. (3.2)) requires further settings to achieve a satisfactory plot.

$$y(x) = \frac{1}{x}. \tag{3.2}$$

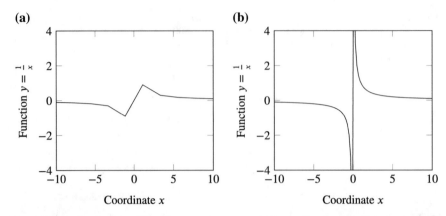

**Fig. 3.11** Plotting of a function with a pole: **a** low number of sampling points (10) and **b** sufficient number of sampling points (100), see Listing 3.19

Plotting with default setting does not produce the desired result as shown in Fig. 3.11a. It is necessary to use the **samples=...** command to specify a sufficient number of sampling points as indicated in Listing 3.19, see Fig. 3.11b for the improved result.

```
1   ...
2   \addplot [domain=-10:10, samples=100]{1/x};
3   ...
```

**Listing 3.19** Plotting of a function under specification of sampling points

The plotting of the cube root for negative and positive arguments requires some special considerations, see Fig. 3.4a. Problems may occur for negative arguments since a complex number is evaluated, see [43]. Thus, the original statement of the cube root, i.e.

$$y(x) = x^{\frac{1}{3}} \tag{3.3}$$

must be transformed to

$$y(x) = \frac{x}{|x|}|x|^{\frac{1}{3}} , \tag{3.4}$$

or alternatively split in different regions:

$$y(x) = \begin{cases} x^{1/3} & \text{for} \quad x > 0 \\ 0 & \text{for} \quad x = 0 \\ -|x|^{1/3} & \text{for} \quad x < 0 \end{cases} . \tag{3.5}$$

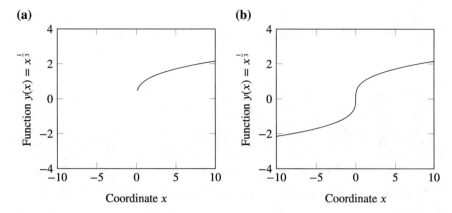

**Fig. 3.12** Plotting of the cube root: **a** attempt based on original equation (3.3) results in missing negative branch and **b** resolution of plotting problem based on Eq. (3.4), see Listing 3.20

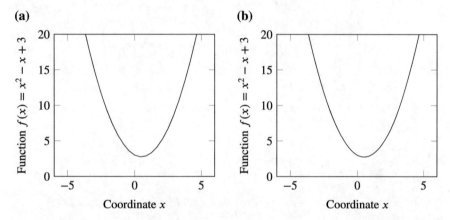

**Fig. 3.13** Graphical representations of functions: **a** original and **b** using the smooth option

For plotting purposes, the approach in Eq. (3.4) might be more convenient and is used in the following Listing 3.20, see Fig. 3.12b for the graphical representation.

```
1   ...
2   \addplot [domain=−10:10, samples=300]{x/abs(x)*abs(x)^(1/3)};
3   ...
```

**Listing 3.20** Plotting of the cube root function

Another possibility to improve the graphical representations of functions is the consideration of the smooth option (option_a), which interpolates smoothly between consecutive data points, see Fig. 3.13.

Let us look in the following on the graphical representation of trigonometric functions. As an example, let us consider the sine and cosine functions:

$$y(x) = \sin(x), \quad y(x) = \cos(x).    \tag{3.6}$$

The trigonometric functions in PGFPLOTS expect the argument in degrees. The conversion of radians to degrees can be easily done with the command deg ( . . . )

```
1    ...
2    \addplot[domain=0:2*pi, samples=100]{sin( deg(x))};
3    ...
```

**Listing 3.21**  Setting to plot a trigonometric function with conversion from radians to degrees

The following listing provides the plots of the sine and cosine functions in the domain $x \in [0, 2\pi]$, see Fig. 3.14.

```
1    \begin{figure}
2    \centering
3         \begin{subfigure}[t]{0.48\textwidth}
4         \captionsetup{labelfont=bf}
5         \caption[]{\hspace*{5cm}}
6              \begin{tikzpicture}
7              \begin{axis}[width=1.00\textwidth,height=0.90\textwidth,
8              xlabel={Coordinate $x$},ylabel={Function $y(x)=\sin(x)$},
9              xmin=0, xmax=6.2832,ymin=-2.0, ymax=2.0,
10             xtick={0,1.5708,3.1416,4.7124,6.2832},
11             xticklabels={0,$\frac{\pi}{2}$,$\pi$,$\frac{3\pi}{2}$,$2\pi$},
12             y label style={at={(axis description cs:0.1,0.5)}}]
13             \addplot[domain=0:2*pi, samples=100]{sin(deg(x))};
14             \draw[] (axis cs:0.0,0) — (axis cs:8.0,0.0);
15             \end{axis}
16             \end{tikzpicture}
17        \end{subfigure}
18        \begin{subfigure}[t]{0.48\textwidth}
19        \captionsetup{labelfont=bf}
20        \caption[]{\hspace*{5cm}}
21             \begin{tikzpicture}
22             \begin{axis}[width=1.00\textwidth,height=0.90\textwidth,
23             xlabel={Coordinate $x$},ylabel={Function $y(x)=\cos(x)$},
24             xmin=0, xmax=6.2832,ymin=-2.0, ymax=2.0,
25             xtick={0,1.5708,3.1416,4.7124,6.2832},
26             xticklabels={0,$\frac{\pi}{2}$,$\pi$,$\frac{3\pi}{2}$,$2\pi$},
27             y label style={at={(axis description cs:0.1,0.5)}}]
28             \addplot[domain=0:2*pi, samples=100]{cos(deg(x))};
29             \draw[] (axis cs:0.0,0) — (axis cs:8.0,0.0);
30             \end{axis}
31             \end{tikzpicture}
32        \end{subfigure}
33   \caption{...}
34   \label{fig:sin_cos}
35   \end{figure}
```

**Listing 3.22**  Plotting of the sine and cosine functions

It can be seen from the Listing 3.22 of Fig. 3.14 that the definition of the locations of the $x$-ticks is a bit cumbersome (see lines 10 and 25). Instead of giving, for example, the numerical value 1.5708, it would be more elegant to simply calculate it as $\frac{\pi}{2}$. This can be achieved by involving the TikZ library calc and the LaTeXpackages tikz and xintexpr. The command \xintdefvar allows to define a constant while the

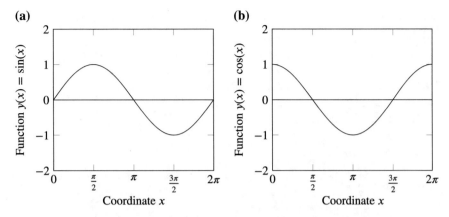

**Fig. 3.14**  Plotting of trigonometric functions: **a** sin(x) and **b** cos(x), see Listing 3.22

environment \xintthefloatexpr ... \relax allows some basic calculations.[1] The main
parts of the modified listing are as follows:

```
1   \documentclass{article}
2
3   \usepackage{tikz}
4   \usetikzlibrary{calc}
5   \usepackage{pgfplots}
6   \usepackage{xintexpr}
7
8   \begin{document}
9
10  ...
11  \begin{tikzpicture}
12      \xintdefvar  Pi:=3.141592653589793;
13      \begin{axis}[width=1.00\textwidth,height=0.90\textwidth,
14      xlabel={Coordinate $x$},ylabel={Function $y(x)=\sin(x)$},
15      xmin=0, xmax=6.2832,ymin=-2.0, ymax=2.0,
16      xtick={0,\xintthefloatexpr Pi/2    \relax,
17              \xintthefloatexpr Pi      \relax,
18              \xintthefloatexpr 3*Pi/2 \relax,
19              \xintthefloatexpr 2*Pi    \relax},
20      xticklabels={0,$\frac{\pi}{2}$,$\pi$,$\frac{3\pi}{2}$,$2\pi$},
21      y label style={at={(axis description cs:0.1,0.5)}}]
22      \addplot[domain=0:2*pi, samples=100]{sin(deg(x))};
23      \draw[] (axis cs:0.0,0) — (axis cs:8.0,0.0);
24  ...
```

**Listing 3.23**  Basic calculations in PGFPLOTS

An alternative way can be based on the LATEXpackage pgfmath, which is normally
automatically loaded with the pgfplots package, see Listing 3.24.

```
1   \documentclass{article}
2
3   \usepackage{pgfplots}
4
5   \begin{document}
6
```

---

[1] The coverage of xintexpr beyond simple math algebra is limited to the square root function, i.e.
sqrt(...).

**Table 3.12** Some available mathematical expressions and the corresponding syntax. For more details, see [39]

| abs($x$) | acos($x$) | asin($x$) | atan($x$) | cos($x$) |
|----------|-----------|-----------|-----------|----------|
| cosh($x$) | cot($x$) | deg($x$) | e | exp($x$) |
| int($x$) | ln($x$) | log10($x$) | log2($x$) | neg($x$) |
| pi | rad($x$) | real($x$) | sec($x$) | sign($x$) |
| sin($x$) | sinh($x$) | sqrt($x$) | tan($x$) | tanh($x$) |

```
 7   ...
 8   \begin{tikzpicture}
 9
10   \pgfmathparse{pi/2}    \pgfmathresult \let\pihalf\pgfmathresult;
11   \pgfmathparse{pi}      \pgfmathresult \let\onepi\pgfmathresult;
12   \pgfmathparse{3*pi/2}  \pgfmathresult \let\threehalfpi\pgfmathresult;
13   \pgfmathparse{2*pi}    \pgfmathresult \let\twopi\pgfmathresult;
14
15   \begin{axis}[width=1.00\textwidth,height=0.90\textwidth,
16   xlabel={Coordinate $x$},ylabel={Function $y(x)=\sin(x)$},
17   xmin=0, xmax=\twopi,ymin=-2.0, ymax=2.0,
18   xtick={0,\pihalf,\onepi,\threehalfpi,\twopi},
19   xticklabels={0,$\frac{\pi}{2}$,$\pi$,$\frac{3\pi}{2}$,$2\pi$},
20   y label style={at={(axis description cs:0.1,0.5)}}]
21   \addplot[domain=0:2*pi, samples=100]{sin(deg(x))};
22   \draw[] (axis cs:0.0,0) -- (axis cs:8.0,0.0);
23   ...
```

**Listing 3.24** Basic calculations in PGFPLOTS based on pgfmath

The command **\pgfmathparse{...}** is quite powerful and allows access to all common mathematical operations and functions, see Part VIII Mathematical and Object-Oriented Engines in [39]. The result of the calculation is stored in **\pgfmathresult** and the construct **\let\...\pgfmathresult** assigns this result to a variable name (\...). After using the absolute function in Listing 3.20 and the sine and cosine functions in Listing 3.22, it might be appropriate to summarizer further, see Table 3.12.

### 3.1.3  Bar Charts

Another typical representation of numerical data is the bar chart. Figure 3.15 shows a simple representation of a bar chart where a factor is plotted for three different materials (i.e., St, Al and Ti). The vertical bar chart is indicated by the axis option **ybar**. It should be noted that a similar approach can be taken for a vertical bar chart (**xbar**).

Attention should be given to the symbolic definition of the $x$-axis. This is indicated by the axis option **symbolic x coords={...}**, see Listing 3.25. The user may use the assigned symbolic values as abscissae values in any further definition.

```
1   \documentclass{article}
2
3   \usepackage{pgfplots}
```

**Fig. 3.15** A simple bar chart, see Listing 3.25

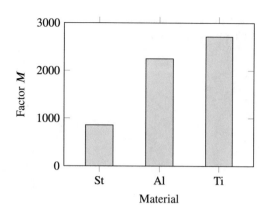

**Table 3.13** Different plot options in PGFPLOTS for bar charts (option_a)

| option_a | Comment |
| --- | --- |
| ybar | This option generates vertical bar plot |
| bar width=7mm | Sets the bar width to 7 mm |
| enlarge x limits={abs=9mm} | Enlarges the distance of the bars to the axes borders to avoid contact |
| symbolic x coords={...,...,...} | Allows to define any string symbols as input coordinates |
| bar shift=0pt | Configures a shift for bars to draw multiple bar plots into the same axis |

```
4
5   \begin{document}
6
7   ...
8   \begin{figure}[h!]
9   \centering
10  \begin{tikzpicture}
11  \begin{axis}[width=0.55\textwidth, height=0.44\textwidth,ybar,bar width=7mm
12              enlarge x limits={abs=9mm},xlabel={Material}, ylabel={Factor $M$},
13              symbolic x coords={St,Al,Ti},xtick={St,Al,Ti},ymin=0, ymax=3000,
14              ytick={0,1000,2000,3000},y label style={at={(axis description cs:
15              0.0,.5)}},,/pgf/number format/1000 sep={}]
16  \addplot[fill=lightgray] coordinates {(St,856.0) (Al,2248.8) (Ti,2711.1)};
17  \end{axis}
18  \end{tikzpicture}
19  \caption{A simple bar chart}
20  \label{fig:pgf_bar_01}
21  \end{figure}
```

**Listing 3.25** A simple bar chart

Specific commands for the definition of bar charts are collected in Table 3.13. These commands relate to the axis specification.

Looking at Fig. 3.15, one may change the numbering of the vertical axis by using an axis multiplier, see Fig. 3.16. This can be achieved with the command scaled y

**Fig. 3.16**  A simple bar chart
with axis multiplier, see
Listing 3.26

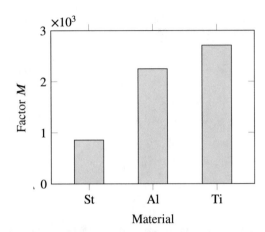

ticks=base 10:-3 which multiplies each number of the vertical axis by $10^{-3} = 0.001$
and introduces the axis multiplier $\times 10^3$, see Listing 3.26.

```
1   \begin{figure}[h!]
2   \centering
3   \begin{tikzpicture}
4   \begin{axis}[ scaled y ticks=base 10:-3, tick scale binop=\times,
5               width=0.55\textwidth, height=0.44\textwidth,ybar,bar width=7mm,
6               enlarge x limits={abs=9mm},xlabel={Material}, ylabel={Factor $M$},
7               symbolic x coords={St,Al,Ti},xtick={St,Al,Ti},ymin=0, ymax=3000,
8               ytick={0,1000,2000,3000},y label style={at={(axis description cs:
9               0.1,.5)}}]
10  %
11  \addplot[fill=lightgray] coordinates {(St,856.0) (Al,2248.8) (Ti,2711.1)};
12  \end{axis}
13  \end{tikzpicture}
14  \caption{A simple bar chart with axis multiplier}
15  \label{fig:pgf_bar_01_multi}       % Give a unique label
16  \end{figure}
```

**Listing 3.26**  A simple bar chart with axis multiplier

A more customized axis multiplier is used in Fig. 3.17 where in addition the location of the multiplier is explicitly specified. The customization is achieved with the
command scaled y ticks=manual:{$...$}{\pgfmathparse{#1...}}: The first argument defines the layout of the multiplier while the second argument defines what is
done with each tick number (represented by the variable #1). The location of the
multiplier if defined by the command every y tick scale label/.style={at={(rel axis
cs:0,1.05)}, anchor = south west, inner sep=1pt}.

```
1   \begin{figure}[h!]
2   \centering
3   \begin{tikzpicture}
4   \begin{axis}[scaled y ticks=manual:{$\times 100$}{\pgfmathparse{#1/100}},
5               tick scale binop=\times, every y tick scale label/.style={at=
6               {(rel axis cs:0,1.05)}, anchor = south west, inner sep=1pt},
7               width=0.55\textwidth, height=0.44\textwidth,ybar,bar width=7mm,
8               enlarge x limits={abs=9mm},xlabel={Material}, ylabel={Factor $M$},
9               symbolic x coords={St,Al,Ti},xtick={St,Al,Ti},ymin=0, ymax=3000,
10              ytick={0,1000,2000,3000},y label style={at={(axis description cs:
11              0.1,.5)}}]
```

**Fig. 3.17** A simple bar chart with customized axis multiplier and customized position, see Listing 3.27

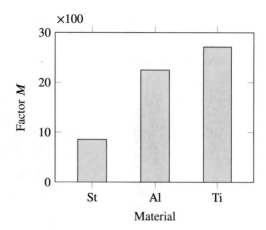

```
12  %
13  \addplot[ fill=lightgray] coordinates {(St,856.0) (Al,2248.8) (Ti,2711.1)};
14  \end{axis}
15  \end{tikzpicture}
16  \caption{A simple bar chart with customized axis multiplier and customized ...}
17  \label{fig:pgf_bar_01_multi_plus}      % Give a unique label
18  \end{figure}
```

**Listing 3.27** A simple bar chart with customized axis multiplier and customized position

A more advanced bar chart is shown in Fig. 3.18 where three values are plotted for each abscissa value. This is achieved by threefold application of the plot command \addplot, see Listing 3.28 for details.

```
1   \documentclass{article}
2
3   \usepackage{pgfplots}
4
5   \begin{document}
6
7   ...
8   \begin{figure}[h!]
9   \centering
10  \begin{tikzpicture}
11  \begin{axis}[width=0.55\textwidth, height=0.6\textwidth,ybar,bar width=7mm,
12              enlarge x limits={abs=14mm},x=3.5cm,,xlabel={Material}, ylabel=
13              {Factor $M$},symbolic x coords={St,Al,Ti},ymin=0,xtick={St,Al,Ti},
14              ymax=3000,ytick={0,1000,2000,3000},y label style={at={(axis
15              description cs:0.00,.5)}},,/pgf/number format/1000 sep={},legend
16              entries={$\sigma_\text{max}$,$u_\text{max}$,$F_\text{max}$},legend
17              style={xshift=-0.0mm,yshift=55.0mm,above of=L1,legend
18              columns=-1,column sep=5mm}]
19  \addplot[area legend, fill=blue!30] coordinates{(St,856) (Al,2248) (Ti,2711)};
20  \addplot[area legend, fill=red!30] coordinates{(St,2548) (Al,2743) (Ti,2277)};
21  \addplot[area legend, fill=green!30] coordinates{(St,653) (Al,1853) (Ti,1084)};
22  \node(L1) at (axis cs:Al,0) {};
23  \end{axis}
24  \end{tikzpicture}
25  \caption{A more advanced bar chart}
26  \label{fig:pgf_bar_01}
27  \end{figure}
```

**Listing 3.28** A more advanced bar chart

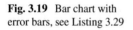

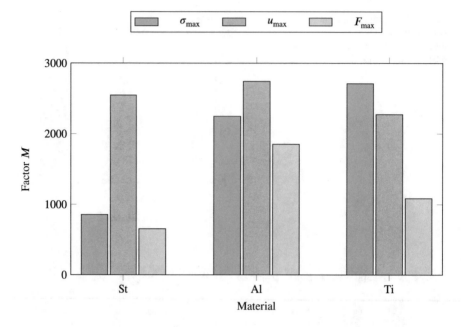

**Fig. 3.18**  A more advanced bar chart, see Listing 3.28

**Fig. 3.19**  Bar chart with
error bars, see Listing 3.29

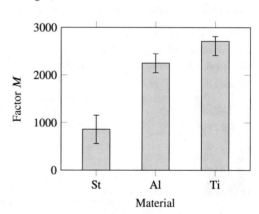

As in the case of single data points, error bars can be introduced in the same way,
see Fig. 3.19 and Listing 3.29.

```
1   \documentclass{article}
2
3   \usepackage{pgfplots}
4
5   \begin{document}
6
7   ...
8   \begin{figure}[h!]
9   \centering
10  \begin{tikzpicture}
```

**Fig. 3.20** Bar chart with
different bar colors, see
Listing 3.30

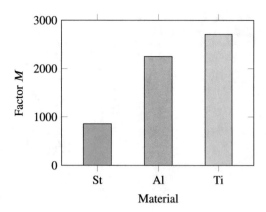

```
11   \begin{axis}[width=0.55\textwidth, height=0.44\textwidth,ybar,bar width=7mm,
12                enlarge x limits={abs=9mm},xlabel={Material}, ylabel=
13                {Factor $M$},symbolic x coords={St,Al,Ti},xtick=
14                {St,Al,Ti},ymin=0, ymax=3000,ytick={0,1000,2000,3000},
15                y label style={at={(axis description cs:0.0,.5)}},
16                /pgf/number format/1000 sep={}]
17   \addplot[fill=lightgray,error bars/.cd,y dir=both,y explicit] coordinates
18        {(St,856.0)+-(0,300) (Al,2248.8)+-(0,200) (Ti,2711.1) +=(0,100)
19        -=(0,300)};
20   \end{axis}
21   \end{tikzpicture}
22   %\end{subfigure}
23   \caption{Bar chart with error bars}
24   \label{fig:pgf_bar_03}
25   \end{figure}
```

**Listing 3.29** Bar chart with error bars

If the bar chart of Fig. 3.15 should be plotted with different colors for each bar,
then each bar must be defined as a single plot and the option **bar shift=0pt** must be
involved, see Fig. 3.20 and Listing 3.30.

```
1    \documentclass{article}
2
3    \usepackage{pgfplots}
4
5    \begin{document}
6
7    ...
8    \begin{figure}[h!]
9    \centering
10   \begin{tikzpicture}
11   \begin{axis}[width=0.55\textwidth, height=0.44\textwidth,ybar,bar width=7mm,
12                enlarge x limits={abs=9mm}, bar shift=0pt,xlabel=
13                {Material}, ylabel={Factor $M$},symbolic x coords=
14                {St,Al,Ti},xtick={St,Al,Ti},ymin=0, ymax=3000,ytick=
15                {0,1000,2000,3000},y label style={at={(axis
16                description cs:0.0,.5)}},,/pgf/number format/1000
17                sep={}]
18   \addplot[fill=blue!30] coordinates {(St,856.0)};
19   \addplot[fill=red!30] coordinates {(Al,2248.8)};
20   \addplot[fill=green!30] coordinates {(Ti,2711.1)};
21   \end{axis}
22   \end{tikzpicture}
23   %\end{subfigure}
```

**Fig. 3.21** Example of box plots, see Listing 3.31

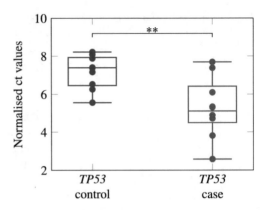

```
24   \caption{Bar chart with different bar colors}
25   \label{fig:pgf_bar_04}
26   \end{figure}
```

**Listing 3.30**  Bar chart with different bar colors

Groups of numerical data are many times represented in descriptive statistics by so-called box plots. The incorporation of the pgfplotslibrary statistics is required for the following example, see Fig. 3.21 and Listing 3.31.

```
1    \documentclass{article}
2
3    \usepackage{pgfplots}
4    \usepgfplotslibrary{statistics}
5
6    \begin{document}
7    ...
8    \begin{figure}[h!]
9    \centering
10           \begin{tikzpicture}
11           \begin{axis}[width=0.55\textwidth, height=0.44\textwidth,ymin=2,
12                        ymax=10,ylabel = {Normalised ct values},
13                        boxplot/draw direction=y,xtick={1,3},
14                        xticklabels={{TP53}\\control, \textit{TP53}\\case},
15                        x tick label style={font=\footnotesize, text width=2cm,
16                        align=center}]
17           \addplot+[mark = *,mark options = {blue},
18                        boxplot prepared={
19                        median=7.395230293,
20                        upper quartile=7.930883408,
21                        lower quartile=6.459313393,
22                        upper whisker=8.21972847,
23                        lower whisker=5.560541153,
24                        every box/.style={thick},
25                        every whisker/.style={thick},
26                        every median/.style={thick},}]
27                        coordinates {
28                                   (1,7.381)
29                                   (1,5.561)
30                                   (1,7.880)
31                                   (1,8.084)
32                                   (1,6.526)
33                                   (1,6.260)
34                                   (1,7.410)
35                                   (1,8.220)
```

```
36                          (1,7.165)};
37              \addplot+[mark = *,mark options = {red},
38                  boxplot prepared={
39                      draw position=3,
40                      median=5.119175911,
41                      upper quartile=6.418040276,
42                      lower quartile=4.489517212,
43                      upper whisker=7.690841675,
44                      lower whisker= 2.576639175,
45                      every box/.style={thick},
46                      every whisker/.style={thick},
47                      every median/.style={thick},}]
48                  coordinates {
49                              (3,4.716)
50                              (3,2.577)
51                              (3,6.096)
52                              (3,3.811)
53                              (3,7.691)
54                              (3,5.343)
55                              (3,4.896)
56                              (3,7.383)
57                              (3,5.314)};
58              \draw[] (axis cs: 1,9.0) — (axis cs: 1,9.2) — (axis cs: 3,9.2)
59                  — (axis cs: 3,9.0);
60              \node[] at (axis cs:2,9.3) {**};
61              \end{axis}
62              \end{tikzpicture}
63  \caption{Example of box plots}
64  \label{fig:pgf_box}
65  \end{figure}
```

**Listing 3.31**  Example of box plots

### *3.1.4  Regression of Data*

Scientific data often requires the evaluation of regression or extrapolation curves to better visualize the trends of single data points.

The functionality of PGFPLOTS is limited to the plotting of *linear* regressions. This requires that the package pgfplotstable is included in the document preamble as shown in the listing below [13]:

```
1  \documentclass[...]{....}
2  \usepackage{pgfplots}
3  \usepackage{pgfplotstable}
4  ...
5  \begin{document}
6  ...
```

**Listing 3.32**  Incorporation of the pgfplotstable package

Let us consider in the following a data table with the file name data_table_2.txt, which is located in a subdirectory ('tables') of the actual working directory. This ASCII file contains two columns which represent six data points $(x_i, y_i)$.

```
x_1 y_1
0   1.0
1   1.4
2   1.4
3   1.4
```

**Fig. 3.22** Linear regression of data points based on $y(x) = a + bx$, see Listing 3.33

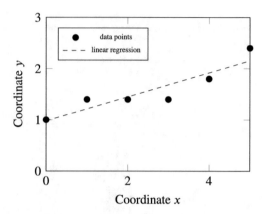

```
        4   1.8
        5   2.4
```

The following Listing 3.33 explains the plotting of the six data points and the corresponding linear regression curve (see also Fig. 3.22). The linear regression itself is defined by the command **create col/linear regression={y=...}.**

```
1   \begin{figure}\centering
2   \pgfplotstableread{tables/data_table_2.txt}\datatable
3       \begin{tikzpicture}
4           \begin{axis}[
5           width=0.55\textwidth, height=0.44\textwidth,
6           xlabel={Coordinate $x$}, ylabel={Coordinate $y$},
7           xmin=0, xmax=5,ymin=0, ymax=3,
8           legend style={at={(axis cs:2.5,2.8)},font=\tiny}]
9           \addplot[only marks,mark=*,color=black]
10              table[x = x_1, y = y_1] from \datatable;
11          \addplot[color=black,mark=none,dashed]
12              table[x=x_1,y={ create col/linear regression={y=y_1}}]
13                              from \datatable;
14          \legend{data points,linear regression}
15          \end{axis}
16      \end{tikzpicture}
17  \end{figure}
```

**Listing 3.33** Linear regression of data points

To plot nonlinear regression curves one can use gnuplot in the background [14]. This freely distributed command-line driven graphing utility is available for all common operating systems (MS Windows, Linux, and MacOS). After installation, it can be easily called within the **\begin{tikzpicture}...\end{tikzpicture}** environment. It provides, in addition, the free parameter fitting of the proposed fitting function.

The LaTeXfile **test.tex** must be compiled from a shell with the following command to produce a PDF file:

pdflatex -shell-escape test

Let us consider in the following a data table with the file name **data_table_3.txt**, which is located in a subdirectory ('tables') of the actual working directory. This ASCII file contains two columns which represent six data points $(x_i, y_i)$.

**Fig. 3.23** Nonlinear
regression of data points
based on $y(x) = a + bx^2$.
Final set of fit parameters:
$a = 1.34324$ and
$b = 0.955283$, see
Listing 3.34

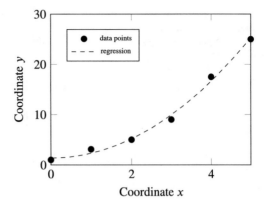

| x_1 | y_1 |
|-----|------|
| 0 | 1.0 |
| 1 | 3.1 |
| 2 | 5.0 |
| 3 | 9.0 |
| 4 | 17.5 |
| 5 | 25.0 |

The following Listing 3.34 explains the plotting of the six data points and the corresponding nonlinear regression curve ($y(x) = a + bx^2$) based on a polynomial (see also Fig. 3.23).

```
1  \begin{figure}\centering
2  \pgfplotstableread{tables/data_table_3.txt}\datatable
3      \begin{tikzpicture}
4      \begin{axis}[width=0.55\textwidth, height=0.44\textwidth,
5      xlabel={Coordinate $x$},ylabel={Data $y$},xmin=0,xmax=5,ymin=0,ymax=30,
6      xmin=0, xmax=5,ymin=0, ymax=30,
7      legend style={at={(axis cs:2.2,27.0)},font=\tiny}]
8      \addplot[only marks,mark=*,color=black] table[x = x_1, y = y_1]
9                                    from \datatable;
10     \addplot[color=black,mark=none,dashed] gnuplot [raw gnuplot]
11         {f(x) = a+b*x^2;a=0.5; b=0.8;fit f(x) 'figs/data_table_3.txt'
12         using 1:2 via a,b; plot [x=0.0:5.0] f(x)};
13     \legend{data points,regression}
14     \end{axis}
15     \end{tikzpicture}
16 \end{figure}
```

**Listing 3.34** Nonlinear regression of data points

The call of gnuplot starts in line 10 of the listing. The given values for $a$ and $b$ are reasonable starting values for the iterative fitting procedure of gnuplot. The statement using 1:2 defines that the first and second column of the text file should be used for the data regression.

Let us now consider a slightly different regression example. The data table with the file name data_table_4.txt is located in a subdirectory ('tables') of the actual working directory. This ASCII file contains two columns which represent 11 data points ($x_i$, $y_i$).

| x_1 | y_1 |
|-------|-------|
| 0.001 | 0.001 |

**Fig. 3.24** Nonlinear regression of data points based on $y(x) = \frac{a}{1+(b/x)^c}$. Final set of fit parameters: $a = 30$, $b = 2.2$, and $c = 3$, see Listing 3.35

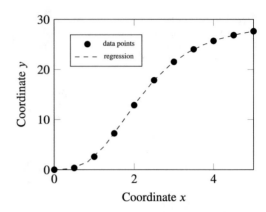

| | |
|---|---|
| 0.5 | 0.348092453 |
| 1.0 | 2.575549451 |
| 1.5 | 7.220280967 |
| 2.0 | 12.87001287 |
| 2.5 | 17.8415103 |
| 3.0 | 21.51508712 |
| 3.5 | 24.03172468 |
| 4.0 | 25.72071589 |
| 4.5 | 26.86125004 |
| 5.0 | 27.64508139 |

The following Listing 3.35 explains the plotting of the 11 data points and the corresponding nonlinear regression curve ($y(x) = \frac{a}{1+(b/x)^c}$) based on a rational function (see also Fig. 3.24).

```
1   \begin{figure}\centering
2   \pgfplotstableread{figs/data_table_4.txt}\datatable
3       \begin{tikzpicture}
4       \begin{axis}[width=0.55\textwidth, height=0.44\textwidth,
5       xlabel={Coordinate $x$},ylabel={Data $y$},xmin=0,xmax=5,ymin=0,ymax=30,
6       legend style={at={(axis cs:2.5,27.0)},font=\tiny}]
7       \addplot[only marks,mark=*,color=black] table[x = x_1, y = y_1]
8                                          from \datatable;
9       \addplot[color=black,mark=none,dashed] gnuplot [raw gnuplot]
10          {f(x) = a/(1+(b/x)^c);a=25; b=1.5; c=2.0; fit f(x)
11          'figs/data_table_4.txt' using 1:2 via a,b,c;
12          plot [x=0.0001:5.0] f(x)};
13      \legend{data points,regression}
14      \end{axis}
15      \end{tikzpicture}
16  \end{figure}
```

**Listing 3.35** Nonlinear regression of data points based on rational function

### 3.1.5   Filling Areas of Functions, Backgrounds and Grid Lines

A graphical representation of plotted functions may be enriched by filling certain areas. The following Listing 3.36 shows how to fill a defined area of the function

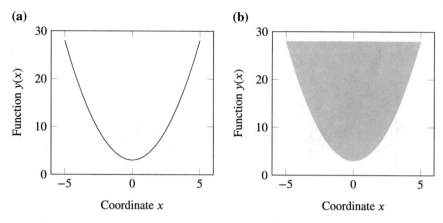

**Fig. 3.25** Filling of areas: **a** original function and **b** filling of the 'inner' domain, see Listing 3.36

being plotted. The dissatisfying issue in Fig. 3.25b is that the gray area is covering the function itself.

```
\begin{figure}
\centering
        \begin{subfigure}[t]{0.48\textwidth}
        \centering
        \captionsetup{labelfont=bf}
        \caption[]{\hspace*{5cm}}
                \begin{tikzpicture}
                \begin{axis}[width=1.00\textwidth, height=0.9\textwidth,
                        xlabel={Coordinate $x$}, ylabel={Function $y(x)$},
                        xmin=-6, xmax=6,ymin=0, ymax=30,]
                \addplot+[mark=none,domain =-5:5,color=black] {x^2 +3};
                \end{axis}
                \end{tikzpicture}
        \end{subfigure}
        \begin{subfigure}[t]{0.48\textwidth}
        \centering
        \captionsetup{labelfont=bf}
        \caption[]{\hspace*{5cm}}
                \begin{tikzpicture}
                \begin{axis}[width=1.00\textwidth, height=0.9\textwidth,
                        xlabel={Coordinate $x$}, ylabel={Function $y(x)$},
                        xmin=-6, xmax=6,ymin=0, ymax=30,]
                \addplot+[mark=none,domain =-5:5,fill,color=gray!50] {x^2 +3};
                \end{axis}
                \end{tikzpicture}
        \end{subfigure}
\caption{...}
\end{figure}
```

**Listing 3.36** Filling of areas: original function and filling of the 'inner' domain

To again make the function visible, the additional plot option **color=back** results in the desired representation, see Fig. 3.26a and Listing 3.37. The command **\closedcycle** allows to fill an area under a graph, see Fig. 3.26b.

```
\begin{figure}
\centering
        \begin{subfigure}[t]{0.48\textwidth}
        \centering
```

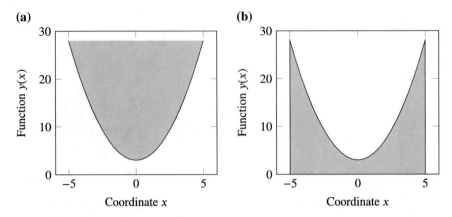

**Fig. 3.26** Filling of areas: **a** refined filling of the 'inner' domain and **b** filling under the graph, see Listing 3.37

```
5    \captionsetup{labelfont=bf}
6    \caption[]{\hspace*{5cm}}
7             \begin{tikzpicture}
8             \begin{axis}[width=1.00\textwidth, height=0.9\textwidth,
9                         xlabel={Coordinate $x$}, ylabel={Function $y(x)$},
10                        xmin=-6, xmax=6,ymin=0, ymax=30,]
11            \addplot+[ color=black,mark=none,domain =-5:5,fill=gray!50]
12                                                   {x^2 +3};
13            \end{axis}
14            \end{tikzpicture}
15    \end{subfigure}
16    \begin{subfigure}[t]{0.48\textwidth}
17    \centering
18    \captionsetup{labelfont=bf}
19    \caption[]{\hspace*{5cm}}
20             \begin{tikzpicture}
21             \begin{axis}[width=1.0\textwidth, height=0.9\textwidth,
22                         xlabel={Coordinate $x$}, ylabel={Function $y(x)$},
23                         xmin=-6, xmax=6,ymin=0, ymax=30,]
24            \addplot+[color=black,mark=none,domain =-5:5,fill=gray!50]
25                                                   {x^2 +3} \closedcycle;
26            \end{axis}
27            \end{tikzpicture}
28    \end{subfigure}
29    \caption{...}
30    \end{figure}
```

**Listing 3.37**  Filling of areas: refined filling of the 'inner' domain and filling under the graph

The filling of an area under a graph requires the application of the \closedcycle command. If only a certain domain under a graph should be filled, double application of the \addplot under different domain definitions may result in the expected representation, see Fig. 3.27a and Listing 3.38. More options are available in the PGFPLOTS library fillbetween, see Fig. 3.27b and Listing 3.38 for an example.

```
1    \documentclass ....
2    ...
3    \usepackage{pgfplots}
4    \usepgfplotslibrary{fillbetween}
5    ...
```

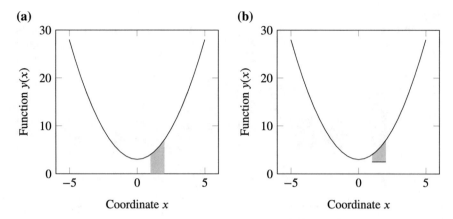

**Fig. 3.27** Filling of areas: **a** domain under a graph and **b** limited domain under a graph, see Listing 3.38

```
6    \begin{document}
7    ...
8    \begin{figure}
9    \centering
10           \begin{subfigure}[t]{0.48\textwidth}
11           \centering
12           \captionsetup{labelfont=bf}
13           \caption[]{\hspace*{5cm}}
14                   \begin{tikzpicture}
15                   \begin{axis}[width=1.00\textwidth, height=0.9\textwidth,
16                               xlabel={Coordinate $x$}, ylabel={Function $y(x)$},
17                               xmin=−6, xmax=6,ymin=0, ymax=30,]
18                   \addplot+[mark=none, fill ,domain=1:2,color=gray!50]
19                                       {x^2 +3} \closedcycle;
20                   \addplot+[mark=none,domain =−5:5,color=black] {x^2 +3};
21                   \end{axis}
22                   \end{tikzpicture}
23           \end{subfigure}
24           \begin{subfigure}[t]{0.48\textwidth}
25           \centering
26           \captionsetup{labelfont=bf}
27           \caption[]{\hspace*{5cm}}
28                   \begin{tikzpicture}
29                   \begin{axis}[width=1.0\textwidth, height=0.9\textwidth,
30                               xlabel={Coordinate $x$}, ylabel={Function $y(x)$},
31                               xmin=−6, xmax=6,ymin=0, ymax=30,]
32                   \addplot+[name path=A,mark=none,domain=1:2,color=black]
33                                       {x^2 +3};
34                   \addplot+[name path=B,mark=none,domain=1:2,color=black] {2.5};
35                   \addplot+[mark=none,domain =−5:5,color=black] {x^2 +3};
36                   \addplot[gray!50] fill between [of = A and B];
37                   \end{axis}
38                   \end{tikzpicture}
39           \end{subfigure}
40   \caption{...}
41   \end{figure}
```

**Listing 3.38** Filling of areas: domain under a graph and limited domain under a graph

The **fill between** option is also useful to fill areas between two functions as shown in Fig. 3.28 and Listing 3.39.

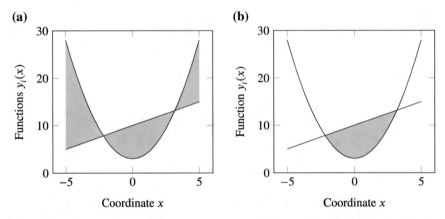

**Fig. 3.28** Filling of areas: **a** between two functions and **b** segment specific, see Listing 3.39

```
1   \begin{figure}
2   \centering
3       \begin{subfigure}[t]{0.48\textwidth}
4       \centering
5       \captionsetup{labelfont=bf}
6       \caption[]{\hspace*{5cm}}
7           \begin{tikzpicture}
8           \begin{axis}[width=1.0\textwidth, height=0.9\textwidth,
9                       xlabel={Coordinate $x$}, ylabel={Functions
10                      $y_i(x)$},xmin=-6, xmax=6,ymin=0, ymax=30,]
11          \addplot+[name path=A,mark=none,domain =-5:5,color=black]
12                      {x^2 +3};
13          \addplot+[name path=B,mark=none,domain =-5:5,color=black]
14                      {10 +x};
15          \addplot[gray!50] fill between[of=A and B];
16          \end{axis}
17          \end{tikzpicture}
18      \end{subfigure}
19      \begin{subfigure}[t]{0.48\textwidth}
20      \centering
21      \captionsetup{labelfont=bf}
22      \caption[]{\hspace*{5cm}}
23          \begin{tikzpicture}
24          \begin{axis}[width=1.0\textwidth, height=0.9\textwidth,
25                      xlabel={Coordinate $x$}, ylabel={Function
26                      $y_i(x)$},xmin=-6, xmax=6,ymin=0, ymax=30,]
27          \addplot+[name path=A,mark=none,domain =-5:5,color=black]
28                      {x^2 +3};
29          \addplot+[name path=B,mark=none,domain =-5:5,color=black]
30                      {10 +x};
31          \addplot[] fill between[of=A and B,split,every segment no
32          0/.style={white},every segment no 1/.style={gray!50},every
33          segment no 2/.style={white}];
34          \end{axis}
35          \end{tikzpicture}
36      \end{subfigure}
37  \caption{...}
38  \end{figure}
```

**Listing 3.39** Filling of areas: between two functions and segment specific

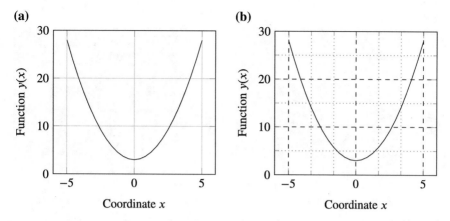

**Fig. 3.29** Grid lines: **a** simple layout and **b** customized layout, see Listing 3.40

Another stylistic feature in data representation is inserting grid lines. This may facilitate the extraction of numerical data from a plotted graph, see Fig. 3.29 and the corresponding Listing 3.40.

```
1   \begin{figure}
2   \centering
3       \begin{subfigure}[t]{0.48\textwidth}
4       \centering
5       \captionsetup{labelfont=bf}
6       \caption[]{\hspace*{5cm}}
7       \begin{tikzpicture}
8       \begin{axis}[width=1.00\textwidth, height=0.9\textwidth,
9                   xlabel={Coordinate $x$}, ylabel={Function $y(x)$},xmin=-6,
10                  xmax=6,ymin=0, ymax=30, grid=major]
11      \addplot+[mark=none,domain =-5:5,color=black] {x^2 +3};
12      \end{axis}
13      \end{tikzpicture}
14      \end{subfigure}
15      \begin{subfigure}[t]{0.48\textwidth}
16      \centering
17      \captionsetup{labelfont=bf}
18      \caption[]{\hspace*{5cm}}
19      \begin{tikzpicture}
20      \begin{axis}[width=1.00\textwidth, height=0.9\textwidth,
21                  xlabel={Coordinate $x$}, ylabel={Function $y(x)$},xmin=-6,
22                  xmax=6,ymin=0, ymax=30,minor x tick num=2,minor y tick num=1,
23                  grid=both,minor grid style={dotted,red},major grid style={dashed,black}]
24      \addplot+[mark=none,domain =-5:5,color=black] {x^2 +3};
25      \end{axis}
26      \end{tikzpicture}
27      \end{subfigure}
28  \caption{Grid lines: \textbf{(a)} simple layout and \textbf{(b)}
29          customized layout}
30  \label{fig:pgf_grid_lines}
31  \end{figure}
```

**Listing 3.40** Grid lines: simple layout and customized layout

The highlighting of an entire graph based on backgrounds is shown in Fig. 3.30 and Listing 3.41.

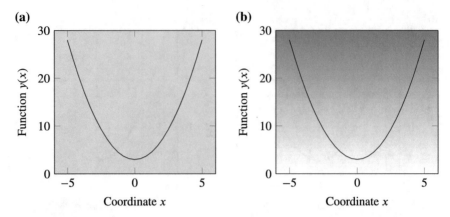

**Fig. 3.30** Background color: **a** simple layout and **b** shaded layout, see Listing 3.41

```
1    \begin{figure}
2    \centering
3        \begin{subfigure}[t]{0.48\textwidth}
4        \centering
5        \captionsetup{labelfont=bf}
6        \caption[]{\hspace*{5cm}}
7        \begin{tikzpicture}
8        \begin{axis}[width=1.00\textwidth, height=0.9\textwidth,
9                    xlabel={Coordinate $x$}, ylabel={Function $y(x)$},xmin=-6,
10                   xmax=6, ymin=0, ymax=30,axis background/.style={fill=
11                   lightgray}]
12       \addplot+[mark=none,domain =-5:5,color=black] {x^2 +3};
13       \end{axis}
14       \end{tikzpicture}
15       \end{subfigure}
16       \begin{subfigure}[t]{0.48\textwidth}
17       \centering
18       \captionsetup{labelfont=bf}
19       \caption[]{\hspace*{5cm}}
20       \begin{tikzpicture}
21       \begin{axis}[width=1.00\textwidth, height=0.9\textwidth,
22                   xlabel={Coordinate $x$}, ylabel={Function $y(x)$},xmin=-6,
23                   xmax=6, ymin=0, ymax=30,axis background/.style={shade,
24                   top color=gray,bottom color=white}]
25       \addplot+[mark=none,domain =-5:5,color=black] {x^2 +3};
26       \end{axis}
27       \end{tikzpicture}
28       \end{subfigure}
29   \caption{Background color: \textbf{(a)} simple layout and \textbf{(b)}
30           shaded layout}
31   \label{fig:pgf_background}
32   \end{figure}
```

**Listing 3.41** Background color: simple layout and shaded layout

**Fig. 3.31** Plotted data points
and popups displaying
coordinates and slope, see
Listing 3.42

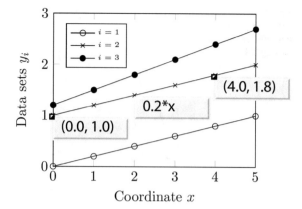

### 3.1.6   Clickable Plots

The pgfplots library clickable[2] allows to create PDFs with clickable graphs. This
enables displaying and extracting the coordinate of a certain marker (simple clicking)
or even the slope between two data points (dragging from one data point to another).
Thus, there is no need for an additional table if the exact values in a graph are of
interest.

Let us come back to Fig. 3.2 where some data from an external file was plotted.
The following Fig. 3.31 shows an interactive version where the coordinates of two
data points are shown and the corresponding slope between these points.

Listing 3.42 highlights the option **clickable coords**, which helps to snap to the
nearest data point.

```
1   \documentclass{article}
2
3   \usepackage{pgfplots}
4   \usepgfplotslibrary{clickable}
5
6   \begin{document}
7
8   \begin{figure}\centering
9   \pgfplotstableread{tables/data_table_1.txt}\datatable
10  \begin{tikzpicture}
11  \begin{axis}[clickable coords,
12           width=0.55\textwidth, height=0.44\textwidth,
13           xlabel={Coordinate $x$}, ylabel={Data sets $y_i$},xmin=0, xmax=5,
14           ymin=0, ymax=3,xtick={0,1,2,3,4,5},legend style={at={(axis cs:1.8,
15           2.8)},font=\tiny}]
16  \addplot[mark=o,color=black] table[x = x_a, y = y_1] from \datatable;
17  \addplot[mark=x,color=black] table[x = x_a, y = y_2] from \datatable;
18  \addplot[mark=*,color=black] table[x = x_a, y = y_3] from \datatable;
19  \legend{$i=1$,$i=2$,$i=3$}
20  \end{axis}
21  \end{tikzpicture}
22  \caption{Plotted data points and popups displaying coordinates and slope}
```

---

[2] This library requires a few style files and definitions which are all provided in the AcroTeX bundle
[38]. In detail, the files eforms.sty, insdljs.sty, taborder.sty, dljscc.def, and eforms.def are required.

**Fig. 3.32**  A simple bar chart
with popup displaying the
bar number and value, see
Listing 3.43

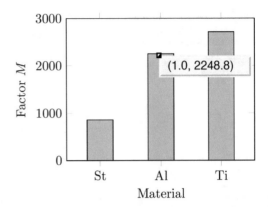

23    \label{fig:fig:clickable_plot_02}
24    \end{figure}

**Listing 3.42**  Plotted data points and popups displaying coordinates and slope

The next example is related to a bar chart, see Fig. 3.32. There are now no single
data points since they are represented by the entire bars. Thus, a user should click
close to the upper end of the bar to display the numerical value. In order to facilitate
this, Listing 3.43 uses the option **snap dist=20**. This increases the distance to snap
to the nearest data point.

```
1   \begin{figure}[h!]
2   \centering
3   \begin{tikzpicture}
4       \begin{axis}[ clickable coords,annot/snap dist=20,width=0.55\textwidth,
5           height=0.44\textwidth,ybar,bar width=7mm,enlarge x limits={abs=9mm},
6           xlabel={Material}, ylabel={Factor $M$},symbolic x coords={St,Al,Ti},
7           xtick={St,Al,Ti},ymin=0, ymax=3000,ytick={0,1000,2000,3000},y label
8           style={at={(axis description cs:0.0,.5)}},,/pgf/number format/
9           1000 sep={}]
10          \addplot[fill=lightgray] coordinates {(St,856.0) (Al,2248.8)
11                                                (Ti,2711.1)};
12      \end{axis}
13  \end{tikzpicture}
14  \caption{A simple bar chart with popup displaying the bar number and value}
15  \label{fig:clickable_plot_02}
16  \end{figure}
```

**Listing 3.43**  A simple bar chart with popups

## 3.1.7 Representation of Functions of Two Variables

### 3.1.7.1 Three-Dimensional Surface Plots

Let us consider in the following, for example, the parameterized representation of an ellipsoid,[3] where the ellipsoid axes coincide with the three Cartesian coordinate axes:

$$x(\theta, \varphi) = 1.0 \times \sin(\theta) \times \cos(\varphi), \tag{3.7}$$

$$y(\theta, \varphi) = 1.0 \times \sin(\theta) \times \sin(\varphi), \tag{3.8}$$

$$z(\theta, \varphi) = 1.5 \times \cos(\theta), \tag{3.9}$$

where the polar angle $\theta$ and the azimuth angle $\varphi$ have the following boundaries:

$$0 \le \theta \le \pi \quad \text{and} \quad 0 \le \varphi \le 2\pi. \tag{3.10}$$

A three-dimensional representation of the surface, i.e. a surface plot, can be obtained based on the command \addplot3[surf], see Listings 3.44 and 3.45. The difference between the two listings is the colored representation of the surface. The first approach is based on the gradual transition of the surface color based on the predefined color map blackwhite (see Fig. 3.33a for the graphical result) whereas the second approach uses a monochrome representation based on the color lightgray[4] (see Fig. 3.33b for the graphical result) and a black surface grid.

```
1   \begin{figure}
2   \centering
3   \begin{tikzpicture}
4   \begin{axis}[xlabel = {$x$}, ylabel = {$y$}, zlabel = {$z$},
5               xmin=-2, xmax=2, ymin=-2, ymax=2, zmin=-1.5, zmax=1.5,
6               view = {60}{30}, domain = 0 : pi, y domain = 0 : 2 * pi,
7               z buffer = sort, unit vector ratio = 1 1,
8               colormap/blackwhite]
9   \addplot3[ surf, samples=30]({sin(deg(x)) * cos(deg(y))}, {1.0 * sin(deg(x)) *
10                             sin(deg(y))}, {1.5 * cos(deg(x))});
11  \end{axis}
12  \end{tikzpicture}
13  \caption{Three-dimensional representation ...}
14  \label{fig:3D_Ellipsoid_1a}
15  \end{figure}
```

**Listing 3.44** Three-dimensional representation of an ellipsoid as a surface plot under consideration of the blackwhite colormap

```
1   \begin{figure}
2   \centering
3   \begin{tikzpicture}
```

---

[3] The implicit form of an ellipsoid has the following standard representation: $\frac{x^2}{a^2} + \frac{y^2}{b^2} + \frac{z^2}{c^2} = 1$. The corresponding parameterized representation reads: $x = a \times \sin(\theta) \times \cos(\varphi)$, $y = b \times \sin(\theta) \times \sin(\varphi)$, and $z = c \times \cos(\theta)$.

[4] See Table A.2 for other colors.

**Fig. 3.33** Three-dimensional representation of an ellipsoid as surface plot: **a** colormap blackwhite and **b** single color (lightgray), see Listings 3.44 and 3.45

**(a)**

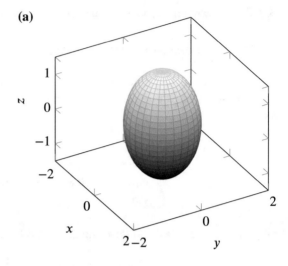

**(b)**

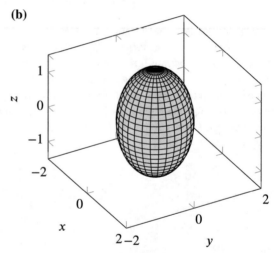

```
4    \begin{axis}[xlabel = {$x$}, ylabel = {$y$}, zlabel = {$z$},
5              xmin=-2, xmax=2, ymin=-2, ymax=2, zmin=-1.5, zmax=1.5,
6              view = {60}{30}, domain = 0 : pi, y domain = 0 : 2 * pi,
7              z buffer = sort, unit vector ratio = 1 1]
8    \addplot3[surf, color=lightgray, faceted color=black, samples=30]
9              ({sin(deg(x)) * cos(deg(y))}, {1.0 * sin(deg(x)) * sin(deg(y))},
10             {1.5 * cos(deg(x))});
11   \end{axis}
12   \end{tikzpicture}
13   \caption{Three-dimensional representation ...}
14   \label{fig:3D_Ellipsoid_1b}
15   \end{figure}
```

**Listing 3.45**  Three-dimensional representation of an ellipsoid as a surface plot with a single color

In case that some parts of the surface should be omitted, the command **restrict z to domain\*=a:b** can be used, see Listings 3.46 and 3.47. In the first example (see also Fig. 3.34a), the top surface is closed and a representation as a solid is obtained. In case that the aim is to showcase the pure surface of the ellipsoid, the command **point meta={zp(\x, \y)>0.499 ? nan : z}** avoids that points with $f(x, y) > 0.499$ get any color assignment (the **nan** stands in this context for 'not a number'), see Fig. 3.34b.

```
1   \begin{figure}
2   \centering
3   \begin{tikzpicture}
4   \begin{axis}[xlabel = {$x$}, ylabel = {$y$}, zlabel = {$z$},
5               xmin=-2, xmax=2, ymin=-2, ymax=2, zmin=-1.5, zmax=1.5,
6               view = {60}{30}, domain = 0 : pi, y domain = 0 : 2 * pi,
7               z buffer = sort,
8               restrict z to domain*=-2.0:0.5, unit vector ratio = 1 1]
9   \addplot3[surf,color=lightgray,faceted color=black,samples=30]({sin(deg(x)) *
10              cos(deg(y))}, {sin(deg(x)) * sin(deg(y))}, {1.5 * cos(deg(x))});
11  \end{axis}
12  \end{tikzpicture}
13  \caption{Representation of an ellipsoid with omitted ...}
14  \label{fig:3D_Ellipsoid_2a}
15  \end{figure}
```

**Listing 3.46** Representation of an ellipsoid with omitted part (closed body)

```
1   \begin{figure}
2   \centering
3   \begin{tikzpicture}
4   \begin{axis}[xlabel = {$x$}, ylabel = {$y$}, zlabel = {$z$},
5               xmin=-2, xmax=2, ymin=-2, ymax=2, zmin=-1.5, zmax=1.5,
6               view = {60}{30}, domain = 0 : pi, y domain = 0 : 2 * pi,
7               z buffer = sort, restrict z to domain*=-2.0:0.5,
8               unit vector ratio = 1 1,
9               declare function = {xp(\x, \y) = sin(deg(\x)) * cos(deg(\y));
10                                  yp(\x, \y) = sin(deg(\x)) * sin(deg(\y));
11                                  zp(\x, \y) = 1.5 * cos(deg(\x))},
12              point meta={zp(\x, \y)>0.499 ? nan : z}]
13  \addplot3[surf,color=lightgray, faceted color=black, samples=30]({xp(x, y)},
14              {yp(x, y)}, {zp(x, y)});
15  \end{axis}
16  \end{tikzpicture}
17  \caption{Representation of an ellipsoid with omitted ...}
18  \label{fig:3D_Ellipsoid_2b}
19  \end{figure}
```

**Listing 3.47** Representation of an ellipsoid with omitted part (pure surface)

It might be worth highlighting the declaration of functions (**xp, yp, zp**) in the options of the axis environment. This allows to easily reuse these functions without retyping the lengthy expression. Variable names are again defined with the (\...) construct (as in Listing 3.24) and Table 3.12 summarizes some of the predefined functions. Within the axis option environment, the construct **\pgfmathparse ... \pgfmathresult** is not required to perform basic mathematics.

**Fig. 3.34** Representation of
an ellipsoid with omitted
part: **a** closed body and **b**
pure surface, see
Listings 3.46 and 3.47

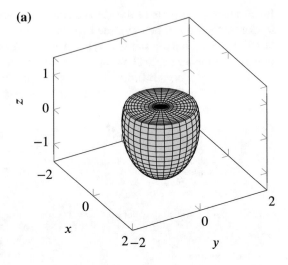

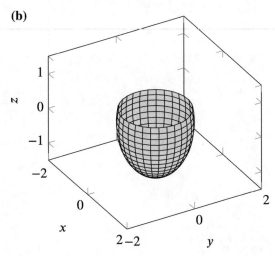

### 3.1.7.2 Contour Plots

A different form of representing a three-dimensional surface can be obtained by
plotting the so-called contour lines.[5] For such a contour line, a function of two
variables takes a constant value, i.e.

$$f(x, y) = \text{const.} \tag{3.11}$$

---

[5] This type of representation is typically known from topographic maps where a contour line represents equal elevation above a given level such as the mean sea level.

**Fig. 3.35** Three-dimensional representation of contour lines for the ellipsoid given in Eq. (3.12) with $a = b = 1$ and $c = 1.5$. The contour lines are given for the constant values $-0.4$, $-0.8$, $-1.2$, and $-1.4$, see Listing 3.48

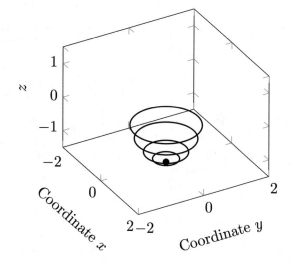

Let us return to the example of an ellipsoid as introduced in Eqs. (3.7)–(3.9). Its representation in explicit form is more advantageous for the representation of contour lines. Rearranging for the $z$-coordinate, we obtain the following general representation:

$$z(x, y) = \pm \sqrt{c^2 \left(1 - \frac{x^2}{a^2} - \frac{y^2}{b^2}\right)}. \tag{3.12}$$

In case that the minimum (i.e., in $z$-direction) should be illustrated, the representation with the minus sign is appropriate. The three-dimensional representation of a few contour lines to illustrate the lower part of the ellipsoid with its minimum at $(0, 0, -1.5)$ for $a = b = 1$ and $c = 1.5$ is shown in Fig. 3.35. The corresponding code is given in Listing 3.48.

```
1  \begin{figure}
2  \centering
3  \begin{tikzpicture}
4  \begin{axis}[xlabel={Coordinate $x$}, ylabel={Coordinate $y$}, zlabel={$z$},
5           xlabel style={sloped like x axis}, ylabel style={sloped like y axis},
6           xmin=-2, xmax=2, ymin=-2, ymax=2, zmin=-1.5, zmax=1.5,
7           view = {60}{30}, unit vector ratio = 1 1,]
8  %
9  \addplot3[contour gnuplot={levels={-0.4,-0.8,-1.2,-1.4}, draw color=black,
10          labels=false}, thick, samples=100, domain=-1.2:1.2]
11          {-1.0*sqrt(2.25*(1-x^2-y^2))};
12 %
13 \addplot3 [only marks,mark=*] coordinates {(0,0,-1.5)};
14 \end{axis}
15 \end{tikzpicture}
16 \caption{...}
17 \label{fig:3D_Ellipsoid_2b}
18 \end{figure}
```

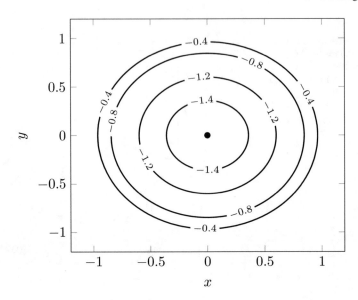

**Fig. 3.36** Contour map of the ellipsoid given in Eq. (3.12) with $a = b = 1$ and $c = 1.5$, see Listing 3.49

**Listing 3.48** Three-dimensional representation of contour lines for the ellipsoid given in Eq. (3.12) with $a = b = 1$ and $c = 1.5$. The contour lines are given for the constant values $-0.4$, $-0.8$, $-1.2$, and $-1.4$

Rotating the three-dimensional diagram given in Fig. 3.35 allows a plane representation, see Fig. 3.36. In such a representation, the different levels can be indicated by labels attached to each contour line. The details of the code are given in Listing 3.49. The command **contour/label distance=...** allows a certain adjustment of the number of labels. A very large number reduces the number of attached labels, in some cases even to a single label.

```
1   \begin{figure}
2   \centering
3   \begin{tikzpicture}
4   \begin{axis}[xlabel = {$x$}, ylabel = {$y$}, zlabel = {$z$},
5            xmin=-1.2, xmax=1.20, ymin=-1.2, ymax=1.2, zmin=-1.5, zmax=1.5,
6            view={0}{90},]
7   %
8   \addplot3[contour gnuplot={levels={-0.4,-0.8,-1.2,-1.4}, draw color=black,
9            contour label style={every node/.append style={text=black}}},
10           thick, samples=100, domain=-1.2:1.2, contour/label distance=100pt]
11           {-1.0*sqrt(2.25*(1-x^2-y^2))};
12  %
13  \addplot3 [only marks,mark=*] coordinates {(0,0,-1.5)};
14  \end{axis}
15  \end{tikzpicture}
16  \caption{...}
17  \label{fig:3D_Ellipsoid_2b}
18  \end{figure}
```

**Listing 3.49** Contour map of the ellipsoid given in Eq. (3.12) with $a = b = 1$ and $c = 1.5$

The formatting of the contour labels can be done in a similar way as in the case of the number formatting of axes labels (see Table 3.2). Listing 3.50 shows, as an example, the required code to change the decimal separator to a comma.

```
7   ...
8   \addplot3[contour gnuplot={levels={−0.4,−0.8,−1.2,−1.4}, draw color=black,
9   contour label style={every node/.append style={text=black,
10  /pgf/number format/.cd,set decimal separator ={,\!}}}},
11  thick, samples=100, domain=−1.2:1.2, contour/label distance=100pt]
12  {−1.0∗sqrt(2.25∗(1−x^2−y^2))};
13  ...
```

**Listing 3.50** Modification of the contour label style: comma as decimal separator

### 3.1.8 Exporting Figures to EPS and PDF files

The previous sections have shown that the source code for TikZ pictures can be directly written in the LATEXdocument. However, it might be required to produce a single EPS or PDF file of a picture which has exactly the size of the bounding box of the picture. Such a file can be used, for example, in other applications. On the other hand, figures based on large data tables may slow down the compilation process of the entire document and it might be advantageous to first create a single EPS or PDF file and then use the includegraphics [...]{...} command to include this figure into the LATEXdocument.

The following Listing 3.51 shows the procedure to generate a single PS file. The commands \usepgfplotslibrary{external} and \tikzexternalize initiates the conversion to an external file while the settings in \tikzset specify the postscript converter.

```
1   \documentclass{article}
2   \usepackage{pgfplots}
3   \usetikzlibrary{pgfplots.groupplots} %optional
4   \usepgfplotslibrary{external}
5   \tikzexternalize[shell escape=−enable−write18]
6
7   \tikzset{external/system call = {latex \tikzexternalcheckshellescape −halt−on
8   −error −interaction=batchmode −jobname "\image" "\texsource"; dvips −o "\image"
9   .ps "\image".dvi}}
10
11  \begin{document}
12
13  \begin{tikzpicture}
14  \begin{axis}[...
15  ...
16  \end{axis}
17  \end{tikzpicture}
18
19  \end{document}
```

**Listing 3.51** Exporting to an external PS file

The LATEXfile **test.tex** must be compiled from the shell with the following command[6] to produce a postscript file called **test-figure0.ps**:

---

[6] Independent from the operating system.

latex -shell-escape test.tex

The following Listing 3.52 shows the procedure to generate a single PDF file.

```
1   \documentclass{article}
2   \usepackage{pgfplots}
3   \usetikzlibrary{pgfplots.groupplots} %optional
4   \usepgfplotslibrary{external}
5   \tikzexternalize
6
7   \begin{document}
8
9   \begin{tikzpicture}
10  \begin{axis}[...
11  ...
12  \end{axis}
13  \end{tikzpicture}
14
15  \end{document}
```

**Listing 3.52**   Exporting to an external PDF file

The L&A;T<sub>E</sub>Xfile test.tex must be compiled from a shell with the following command to produce a PDF file called test-figure0.pdt:

pdflatex -shell-escape test

Another application of Listing 3.52 can be seen for large documents with a lot of pgfplots figures and considerably large compilation times. Using the settings in lines 4 and 5 generates for each pgfplot figure a single PDF file which is automatically read and included during the next compilation without recompiling it. Only if the source code of a figure gets changed, the file is recompiled. Thus, a significant reduction of the compilation time can be achieved.

## 3.2   Figure Generation with TikZ

### 3.2.1   TikZ Matrix Library

The TikZ library matrix allows the advanced representation of equations with a focus on matrices. The following Listing 3.53 explains the definition of matrices based on the \matrix command, see Eq. (3.13) for the result.[7] Each matrix now has its own name, for example, the first equation is called 'mm'. In addition, each matrix cell is now by definition a node, which can be addressed based on the matrix name and two consecutive numbers (the first refers to the row and the second to the column).[8] The option baseline=... is used for the vertical alignment of the matrix in the equation environment.

---

[7] This example is taken from the context of computational mechanics, see [27].

[8] In case of an empty cell, one may add the option nodes in empty cells as an additional argument of the matrix command. This allows to address an empty cell.

```
1   \documentclass{article}
2
3   \usepackage{amsmath}
4
5   \usepackage{pgfplots}
6   \usetikzlibrary{matrix,fit,calc}
7
8   \begin{equation}
9   \cfrac{EA}{L}\times
10  \begin{tikzpicture}[baseline=($(mm-1-1)!.85!(mm-2-1)$)]
11  \matrix (mm) [matrix of math nodes,left delimiter=\lbrack,
12  right delimiter=\rbrack,inner sep=2.0pt,outer sep=0.0pt]
13  {
14  K_{11} & K_{12} \\
15  K_{21} & K_{22} \\
16  };
17  \end{tikzpicture}
18  \begin{tikzpicture}[baseline=($(uu-1-1)!.85!(uu-2-1)$)]
19  \matrix (uu) [matrix of math nodes,left delimiter=\lbrack,
20  right delimiter=\rbrack,inner sep=3.0pt,outer sep=0.0pt]
21  {
22  u_{1x}\\
23  u_{2x}\\
24  };
25  \end{tikzpicture}
26  =
27  \begin{tikzpicture}[baseline=($(nn-1-1)!.85!(nn-2-1)$)]
28  \matrix (nn) [matrix of math nodes,left delimiter=\lbrack,
29  right delimiter=\rbrack,inner sep=2.0pt,outer sep=0.0pt,draw]
30  {
31  F_{1x}\\
32  F_{2x}\\
33  };
34  \end{tikzpicture}
35  \end{equation}
```

**Listing 3.53** Simple matrix equation

$$\frac{EA}{L} \times \begin{bmatrix} K_{11} & K_{12} \\ K_{21} & K_{22} \end{bmatrix} \begin{bmatrix} u_{1x} \\ u_{2x} \end{bmatrix} = \begin{bmatrix} F_{1x} \\ F_{2x} \end{bmatrix} \tag{3.13}$$

A more advanced example is as shown in Listing 3.54, see Eq. (3.15) for the result. The nodes of the matrix are now used to define the square brackets of the first matrix. This is achieved in a new node definition. In addition, the column matrix on the right-hand side is defined based on the bmatrix environment. Thus, mixing of different matrix environments is possible.

```
1   \documentclass{article}
2
3   \usepackage{amsmath}
4
5   \usepackage{pgfplots}
6   \usetikzlibrary{matrix,fit,calc}
7
8   \begin{equation}
9   \cfrac{EA}{L}\times
10  \begin{tikzpicture}[baseline=($(m-1-1)!.85!(m-3-1)$)]
11  \matrix (m) [matrix of math nodes]
12  {
13  u_1  & u_2  &[3mm] \\
14  K_{11} & K_{12} & u_1 \\
```

```
15   K_{21} & K_{22} & u_2   \\
16   };
17   \ node[left delimiter=\lbrack,inner sep=0,fit={(m-2-1)(m-3-1)}]{};
18   \ node[right delimiter=\rbrack,inner sep=0.0,fit={(m-2-2)(m-3-2)}]{};
19   \end{tikzpicture}
20   =
21   \begin{bmatrix}
22   F_1\\
23   F_2
24   \end{bmatrix}
25   \end{equation}
```

**Listing 3.54** Matrix equation with annotation over and beside matrix and different definitions of matrices

$$\frac{EA}{L} \times \begin{bmatrix} u_1 & u_2 \\ K_{11} & K_{12} \\ K_{21} & K_{22} \end{bmatrix} \begin{matrix} u_1 \\ u_2 \end{matrix} = \begin{bmatrix} F_1 \\ F_2 \end{bmatrix} \tag{3.14}$$

Listing 3.55 and Eq. (3.15) shows the same example. However, the right-hand column matrix is now defined in a tikzpicture environment.

```
1    \documentclass{article}
2
3    \usepackage{amsmath}
4
5    \usepackage{pgfplots}
6    \usetikzlibrary{matrix,fit,calc}
7
8    \begin{equation}\let\algoequation=Y
9    \renewcommand{\theequation}{\thesection.\arabic{equation}}
10   \setcounter{section}{15}
11   \mbox{\boldmath$K$}=\cfrac{EA}{L}\times
12   \begin{tikzpicture}[baseline=($(m-1-1)!.85!(m-3-1)$)]
13   \matrix (m) [matrix of math nodes]
14   {
15   u_1   & u_2   &[3mm] \\
16   K_{11} & K_{12} & u_1   \\
17   K_{21} & K_{22} & u_2   \\
18   };
19   \node[left delimiter=\lbrack,inner sep=0,fit={(m-2-1)(m-3-1)}]{};
20   \node[right delimiter=\rbrack,inner sep=0.0,fit={(m-2-2)(m-3-2)}]{};
21   \end{tikzpicture}
22   =
23   \begin{tikzpicture}[baseline=($(n-1-1)!.8!(n-2-1)$)]
24   \matrix (n) [matrix of math nodes,left delimiter=\lbrack,
25   right delimiter=\rbrack,inner sep=2.0pt,outer sep=0.0pt,]
26   {
27   F_{1}\\
28   F_{2}\\
29   };
30   \end{tikzpicture}
31   \end{equation}
```

**Listing 3.55** Matrix equation with annotation over and beside matrix and uniform definition of matrices based on tikzpicture environment

$$K = \frac{EA}{L} \times \begin{bmatrix} \overset{u_1 \quad u_2}{\begin{matrix} K_{11} & K_{12} \\ K_{21} & K_{22} \end{matrix}} \end{bmatrix} \begin{matrix} u_1 \\ u_2 \end{matrix} = \begin{bmatrix} F_1 \\ F_2 \end{bmatrix} \tag{3.15}$$

For some applications, it might be useful to indicate that some matrix rows and columns can be canceled from the system of equations. If these lines should extend over several matrices, i.e. several tikzpicture environments, the tikzpicture options 'remember picture' and 'overlay' should be used, see Listing 3.56 and Eq. (3.16). To extend the horizontal and vertical lines a bit over the indicated nodes (ma... and na...), a 7 and 3 mm offset has been introduced in the coordinate definitions in lines 30 and 31.

```
1   \documentclass{article}
2
3   \usepackage{amsmath}
4
5   \usepackage{pgfplots}
6   \usetikzlibrary{matrix,fit,calc}
7
8   \begin{equation}
9   \mbox{\boldmath$K$}=\cfrac{EA}{L}\times
10  \begin{tikzpicture}[remember picture,baseline=($(ma-1-1)!.85!(ma-3-1)$)]
11  \matrix (ma) [matrix of math nodes]
12  {
13  u_1 & u_2 &[3mm]  \\
14  K_{11} & K_{12} & u_1 \\
15  K_{21} & K_{22} & u_2 \\
16  };
17  \node[color=red,left delimiter=\lbrack,inner sep=0,fit={(ma-2-1)(ma-3-1)}]{};
18  \node[red,right delimiter=\rbrack,inner sep=0.0,fit={(ma-2-2)(ma-3-2)}]{};
19  \end{tikzpicture}
20  =
21  \begin{tikzpicture}[remember picture,baseline=($(na-1-1)!.8!(na-2-1)$)]
22  \matrix (na) [matrix of math nodes,left delimiter=\lbrack,
23  right delimiter=\rbrack,inner sep=2.0pt,outer sep=0.0pt,]
24  {
25  F_{1}\\
26  F_{2}\\
27  };
28  \end{tikzpicture}
29  \begin{tikzpicture}[remember picture,overlay]
30  \draw[red,thick]  ($(ma-2-1)-(7mm0)$) -- ($(na-1-1)+(7mm0)$);
31  \draw[red,thick]  ($(ma-1-1)+(0,3mm)$) -- ($(ma-3-1)-(0,3mm)$);
32  \end{tikzpicture}
33  \end{equation}
```

**Listing 3.56** Symbolic cancelling of a row and column of a matrix

$$K = \frac{EA}{L} \times \begin{bmatrix} \overset{u_1 \quad u_2}{\begin{matrix} K_{11} & K_{12} \\ K_{21} & K_{22} \end{matrix}} \end{bmatrix} \begin{matrix} u_1 \\ u_2 \end{matrix} = \begin{bmatrix} F_1 \\ F_2 \end{bmatrix} \tag{3.16}$$

The length adjustment of the red horizontal line can be also achieved, for example, with the modified command \draw[red,thick] ([xshift=-5mm]m-2-1.west) -- (n-1-1.east);.

The following example shows how a certain block of a matrix can be highlighted, see Listing 3.57 and Eq. (3.17).

```
\documentclass{article}

\usepackage{amsmath}

\usepackage{pgfplots}
\usetikzlibrary{matrix,fit,calc,backgrounds}

\begin{equation}
\mbox{\boldmath$K$}=\cfrac{EA}{L}\times
\begin{tikzpicture}[remember picture,baseline=($(p-1-1)!.6!(p-3-1)$)]
\matrix (p) [matrix of math nodes,left delimiter=\lbrack,
right delimiter=\rbrack,inner sep=2.0pt,outer sep=0.0pt,]
{
K_{11} & K_{12} & K_{13}\\
K_{21} & K_{22} & K_{23}\\
K_{31} & K_{32} & K_{33}\\
};
\begin{pgfonlayer}{background}
\draw[color=lightgray,fill=lightgray,opacity=0.9] (p-2-1.south west) rectangle
(p-1-2.north east);
\end{pgfonlayer}
\end{tikzpicture}
\end{equation}
```

**Listing 3.57**  Highlighling of a block of matrix elements

$$K = \frac{EA}{L} \times \begin{bmatrix} K_{11} & K_{12} & K_{13} \\ K_{21} & K_{22} & K_{23} \\ K_{31} & K_{32} & K_{33} \end{bmatrix} \tag{3.17}$$

Highlighting of single matrix elements is shown in Listing 3.58 and Eq. (3.18).

```
\documentclass{article}

\usepackage{amsmath}

\usepackage{pgfplots}
\usetikzlibrary{matrix,fit,calc,backgrounds}

\begin{equation}
\mbox{\boldmath$K$}=\cfrac{EA}{L}\times
\begin{tikzpicture}[remember picture,baseline=($(pp-1-1)!.6!(pp-3-1)$)]
\matrix (pp) [matrix of math nodes,left delimiter=\lbrack,
right delimiter=\rbrack,inner sep=2.0pt,outer sep=0.0pt,]
{
|[red]|K_{11} & K_{12} & K_{13}\\
K_{21} & K_{22} & K_{23}\\
K_{31} & K_{32} & K_{33}\\
};
\begin{pgfonlayer}{background}
\draw[color=lightgray,fill=lightgray,opacity=0.9] (pp-3-3.south west) rectangle
(pp-3-3.north east);
\end{pgfonlayer}
\end{tikzpicture}
\end{equation}
```

**Listing 3.58**  Highlighling of single matrix elements

$$K = \frac{EA}{L} \times \begin{bmatrix} K_{11} & K_{12} & K_{13} \\ K_{21} & K_{22} & K_{23} \\ K_{31} & K_{32} & K_{33} \end{bmatrix} \tag{3.18}$$

Indicating different blocks in a matrix based on separation lines is shown in Listing 3.59 and Eq. (3.19).

```
1   \documentclass{article}
2
3   \usepackage{amsmath}
4
5   \usepackage{pgfplots}
6   \usetikzlibrary{matrix,fit,calc}
7   ...
8   \begin{equation}
9   \mbox{\boldmath$K$}=\cfrac{EA}{L}\times
10  \begin{tikzpicture}[remember picture,baseline=($(ppp-1-1)!.6!(ppp-3-1)$)]
11  \matrix (ppp) [matrix of math nodes,left delimiter=\lbrack,
12  right delimiter=\rbrack,inner sep=2.0pt,outer sep=0.0pt,]
13  {
14  K_{11} & K_{12} & K_{13}\\
15  K_{21} & K_{22} & K_{23}\\
16  K_{31} & K_{32} & K_{33}\\
17  };
18  \draw[dashed] (ppp-2-1.south west) -- (ppp-2-3.south east);
19  \draw[dashed] (ppp-1-2.north east) -- (ppp-3-2.south east);
20  \end{tikzpicture}
21  \end{equation}
```

**Listing 3.59**   Indicating different blocks of a matrix

$$K = \frac{EA}{L} \times \begin{bmatrix} K_{11} & K_{12} & K_{13} \\ K_{21} & K_{22} & K_{23} \\ K_{31} & K_{32} & K_{33} \end{bmatrix} \tag{3.19}$$

The following examples show the application of horizontal brackets in conjunction with matrices. In Eq. (3.20), the curly bracket spans under a matrix and its implementation is shown in Listing 3.60

$$\frac{EA}{L} \times \underbrace{\begin{bmatrix} K_{11} & K_{12} \\ K_{21} & K_{22} \end{bmatrix}}_{K} \begin{bmatrix} u_{1x} \\ u_{2x} \end{bmatrix} = \underbrace{\begin{bmatrix} F_{1x} \\ F_{2x} \end{bmatrix}}_{F} \tag{3.20}$$

```
1   \documentclass{article}
2
3   \usepackage{amsmath}
4   \usepackage{pgfplots}
5   \usetikzlibrary{matrix,fit,calc}
6   ...
7   \begin{equation}\label{eq:tikz_matrix_deli_1}
8   \cfrac{EA}{L}\times
9   \begin{tikzpicture}[baseline=($(ma-1-1)!.85!(ma-2-1)$)]
```

```
10   \matrix (ma) [matrix of math nodes,left delimiter=\lbrack,right delimiter
11   =\rbrack,inner sep=2.0pt,outer sep=0.0pt]
12   {
13   K_{11} & K_{12} \\
14   K_{21} & K_{22} \\
15   };
16
17   \node[below delimiter=\},fit={($(ma-2-1.south)-(6mm,0)$)($(ma-2-2.south)+
18   (6mm,0)$)}] (del-left) {};
19   \node[below=10pt] at (del-left.south) {\;$\mbox{\boldmath$K$}$};
20   \end{tikzpicture}
21   \begin{tikzpicture}[baseline=($(ua-1-1)!.85!(ua-2-1)$)]
22   \matrix (ua) [matrix of math nodes,left delimiter=\lbrack,right delimiter
23   =\rbrack,inner sep=3.0pt,outer sep=0.0pt]
24   {
25   u_{1x}\\
26   u_{2x}\\
27   };
28   \end{tikzpicture}
29   =
30   \begin{tikzpicture}[baseline=($(na-1-1)!.85!(na-2-1)$)]
31   \matrix (na) [matrix of math nodes,left delimiter=\lbrack,right delimiter
32   =\rbrack,inner sep=2.0pt,outer sep=0.0pt,draw]
33   {
34   F_{1x}\\
35   F_{2x}\\
36   };
37   \node[below delimiter=\},fit={($(na-2-1.south)-(6mm,0)$)($(na-2-1.south)
38   +(6mm,0)$)}] (del-left-F) {};
39   \node[below=10pt] at (del-left-F.south) {$\mbox{\;\boldmath$F$}$};
40   \end{tikzpicture}
41   \end{equation}
```

**Listing 3.60**  Matrices and bracket in the same environment

If the curly bracket in Eq. (3.20) should also comprise the fraction in front of the matrix, a sophisticated solution with separate Ti*k*Z environments and the overlay technique should be used, see Eq. (3.21) and the corresponding Listing 3.61.

$$\underbrace{\frac{EA}{L} \times \begin{bmatrix} K_{11} & K_{12} \\ K_{21} & K_{22} \end{bmatrix}}_{K} \begin{bmatrix} u_{1x} \\ u_{2x} \end{bmatrix} = \begin{bmatrix} \begin{bmatrix} F_{1x} \\ F_{2x} \end{bmatrix} \end{bmatrix} \tag{3.21}$$

```
1    \documentclass{article}
2
3    \usepackage{amsmath}
4    \usepackage{pgfplots}
5    \usetikzlibrary{matrix,fit,calc}
6    ...
7    \begin{equation}
8    \cfrac{EA}{L}\times
9    \begin{tikzpicture}[remember picture,baseline=($(ma-1-1)!.85!(ma-2-1)$)]
10   \matrix (ma) [matrix of math nodes,left delimiter=\lbrack,right delimiter
11   =\rbrack,inner sep=2.0pt,outer sep=0.0pt,]
12   {
13   K_{11} & K_{12} \\
14   K_{21} & K_{22} \\
15   };
16   \end{tikzpicture}
```

```
17   %
18   \begin{tikzpicture}[remember picture,baseline=($(ua-1-1)!.85!(ua-2-1)$)]
19   \matrix (ua) [matrix of math nodes,left delimiter=\lbrack,right delimiter
20   =\rbrack,inner sep=3.0pt,outer sep=0.0pt]
21   {
22   u_{1x}\\
23   u_{2x}\\
24   };
25   \end{tikzpicture}
26   =
27   \begin{tikzpicture}[remember picture,baseline=($(na-1-1)!.85!(na-2-1)$)]
28   \matrix (na) [matrix of math nodes,left delimiter=\lbrack,right delimiter
29   =\rbrack,inner sep=2.0pt,outer sep=0.0pt,draw]
30   {
31   F_{1x}\\
32   F_{2x}\\
33   };
34   \end{tikzpicture}
35   \begin{tikzpicture}[remember picture,overlay]
36   \node[below delimiter=\}, fit={($(ma-2-1.south)-(17mm,0)$)($(ma-2-2.south)
37   +(6mm,0)$)}] (del-left) {};
38   \node[below=10pt] at (del-left.south) {$\mbox{\boldmath$K$}$};
39   \end{tikzpicture}
40   \end{equation}
```

**Listing 3.61** Matrices and bracket in different environments

### 3.2.2 Flowcharts

A flowchart is a specific diagram that illustrates an algorithm, workflow or process, showing the consecutive steps as boxes of various shapes, and their order by connecting them with arrows. There is a LATEX package called flowchart, which provides a set of 'traditional' flowchart element shapes, see [31]. Nevertheless, we will show in the following the generation of flowcharts based on pure TikZ commands and some libraries, i.e. geometric shapes and arrows. This approach allows an easier generation of new shapes and other features, see [33].

The example in Fig. 3.37 shows a simple flowchart. Basic shapes are first defined in the tikzpicture environment. Then we define the specific boxes, which are connected by arrows in the last part of the environment (see Listing 3.62).

```
1    \documentclass{article}
2
3    \usepackage{tikz}
4    \usetikzlibrary{shapes.geometric, arrows}
5    ...
6    \begin{figure}
7    \centering
8    \begin{tikzpicture}[node distance=2cm]
9        % Definitions of Shapes
10       \tikzstyle{startstop} = [rectangle, rounded corners, minimum width=3cm,
11               minimum height=1cm, align=center, draw=black, fill=lightgray]
12       \tikzstyle{io} = [trapezium, trapezium left angle=70, trapezium right
13               angle=110, minimum width=3cm, minimum height=1cm, align=center,
14               draw=black, fill=lightgray]
15       \tikzstyle{process} = [rectangle, minimum width=3cm, minimum height=1cm,
16               align=center, draw=black, fill=lightgray]
17       \tikzstyle{decision} = [diamond, minimum width=3cm, minimum height=1cm,
```

**Fig. 3.37** Example of a
simple flowchart, see
Listing 3.62

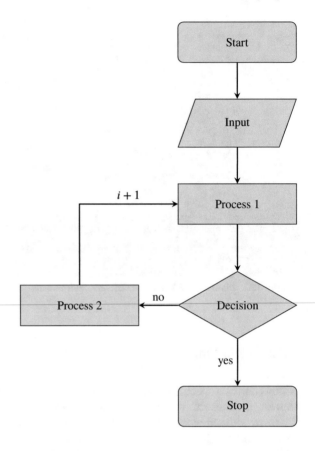

```
18              align=center, draw=black, fill=lightgray]
19        \tikzstyle{arrow} = [thick,->,>=stealth]
20    % Boxes
21        \node (start) [startstop] {Start};
22        \node (in1) [io, below of=start] {Input};
23        \node (pro1) [process, below of=in1] {Process 1};
24        \node (dec1) [decision, below of=pro1,yshift=-0.5cm] {Decision};
25        \node (pro2) [process, left of=dec1, xshift=-2cm] {Process 2};
26        \node (stop) [startstop, below of=dec1, yshift=-0.5cm] {Stop};
27    % Arrows
28        \draw [arrow] (start) -- (in1);
29        \draw [arrow] (in1) -- (pro1);
30        \draw [arrow] (pro1) -- (dec1);
31        \draw [arrow] (dec1) -- node[anchor=south] {no} (pro2);
32        \draw [arrow] (dec1) -- node[anchor=east] {yes} (stop);
33        \draw [arrow] (pro2) |- node[pos=0.75,anchor=south] {$i+1$} (pro1);
34    \end{tikzpicture}
35    \caption{Example of a simple flowchart}
36    \label{fig:simple_flow}
37    \end{figure}
```

**Listing 3.62** Example of a simple flowchart

Advanced formatting of the text in a node, i.e. line breaks and text alignment, can be achieved by introducing a tabular environment, see Listing 3.63 and the corresponding Fig. 3.38.

```
1   \begin{figure}
2   \centering
3   \begin{tikzpicture}[node distance=2cm]
4       % Definitions of Shapes
5       \tikzstyle{startstop} = [rectangle, rounded corners, minimum width=3cm,
6                   minimum height=1cm, align=center, draw=black, fill=lightgray]
7       \tikzstyle{io} = [trapezium, trapezium left angle=70, trapezium right
8                   angle=110, minimum width=3cm, minimum height=1cm, align=center,
9                   draw=black, fill=lightgray]
10      \tikzstyle{process} = [rectangle, minimum width=3cm, minimum height=1cm,
11                  align=center, draw=black, fill=white]
12      \tikzstyle{decision} = [diamond, minimum width=3cm, minimum height=1cm,
13                  align=center, draw=black, fill=lightgray]
14      \tikzstyle{arrow} = [thick,->,>=stealth]
15      % Boxes
16      \node (in1) [io] {
17              \begin{tabular}{l}
18                  \multicolumn{1}{c}{Input:}\\
19                  \multicolumn{1}{c}{$\bullet$ Definition of Geometry:}\\
20                  coordinates \textsf{(ncoor)},\\
21                  cross section \textsf{(A)},\\
22                  second moment of area \textsf{(Izz)}\\[0.6ex]
23                  \multicolumn{1}{c}{$\bullet$ Definition of Material:}\\
24                  Young's modulus \textsf{(Em)}
25              \end{tabular}};
26      \node (pro1) [process, below of=in1, yshift=-0.5cm] {Maxima Module:\\
27              \textsf{K\_el\_gen\_Beam\_xy(ncoor, Em, A, Izz)}};
28      \node (out1) [io, below of=pro1, yshift=0.0cm] {Output:\\
29              stiffness matrix \textsf{(Ke)}};
30      % Arrows
31      \draw [arrow] (in1) -- (pro1);
32      \draw [arrow] (pro1) -- (out1);
33  \end{tikzpicture}
34  \caption{Example of a flowchart with an advanced textbox}
35  \label{fig:fig:simple_flow}
36  \end{figure}
```

**Listing 3.63**  Example of a flowchart with an advanced textbox

### 3.2.3  Remarks on Schematic Drawings

The basic length unit in TikZ is 1 cm. This means that the command \draw (0,0)- -(5,0) draws a horizontal line of length 5 cm. The same could be obviously achieved by \draw (0cm,0cm)- -(5cm,0cm) but the first version is of course shorter. The user may also explicitly specify other units such as 'mm' or 'pt'. If a user intents to change the basic length unit, one can set the $x$- and $y$-vector as shown in Listing 3.64.

```
1   ...
2   \begin{tikzpicture}[x=3cm,y=2.5cm]
3   ...
```

**Listing 3.64**  Change of length unit in TikZ

**Fig. 3.38** Example of a
flowchart with an advanced
textbox, see Listing 3.63

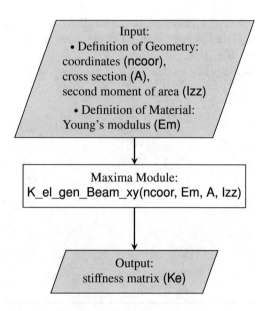

This would mean that the command \draw (0,0)- -(5,0) would sketch a line of length
15 cm. The vector definition in Listing 3.64 is hereby a short form for the complete
vector definition as shown in Listing 3.65

```
1  ...
2  \begin{tikzpicture}[x={3cm,0cm},y={0cm,2.5cm}]
3  ...
```

**Listing 3.65**  Change of length unit in TikZ based on full vector defintion

Some users may prefer to first define the outer dimensions of a picture and then
fill it with elements. Let us recall that the use of pgfplots allowed to exactly define
the length of the axes with width=... and height=.... Such an option is not a priori
available in TikZ. However, the user may define invisible paths in the x- and y-
direction to define the sketching space, see Listing 3.66. This also defines the origin
of the coordinate system in the lower left corner of the sketching area. The lower left
and upper right corner of the sketching plane is indicated by circles in Fig. 3.39.

```
1   \begin{figure}[h!]
2   \centering
3   \begin{tikzpicture}[]
4   %
5       \path (0cm,0cm) --- (0.55\textwidth, 0cm);
6       \path (0cm,0cm) --- (0cm, 0.44\textwidth);
7   %
8       \fill [lightgray] (0,0) circle (5pt);
9       \fill [white] (0.55\textwidth,0.44\textwidth) circle (5pt);
10  %
11  \end{tikzpicture}
12  \caption{Definition the dimensions of the 'sketching plane'}
13  \end{figure}
```

**Listing 3.66**  Definition of the dimensions of the 'sketching plane'

**Fig. 3.39**  Definition of the
dimensions of the 'sketching
plane', see Listing 3.66

Another useful feature with pgfplots was the use of normalized coordinates, i.e. (rel
axis cs:x,y) in which the lower left corner of the graph is (0,0) and upper right corner
is (1,1). Thereby it holds $0 \leq x \leq 1$ and $0 \leq y \leq 1$ for the relative coordinate system.
A similar feature can be achieved in a pure TikZ environment by setting the $x$- and
$y$-vector to the size of the 'sketching plane', see Listing 3.67 and the corresponding
Fig. 3.40.

```
\begin{figure}[h!]
\centering
\begin{tikzpicture}[x=0.55\textwidth,y=0.44\textwidth]
%
        \path (0cm,0cm) ――  (0.55\textwidth, 0cm);
        \path (0cm,0cm) ――  (0cm, 0.44\textwidth);
%
        \draw [ultra thin] (0,0) rectangle (0.55\textwidth,0.44\textwidth);
%
\end{tikzpicture}
\caption{'Sketching plane' with $0 \leq x \leq 1$ and $0 \leq y \leq 1$}
\label{fig:TikZ_define_size_sketchingplane_norm}
\end{figure}
```

**Listing 3.67**  'Sketching plane' with $0 \leq x \leq 1$ and $0 \leq y \leq 1$

On a side note, the actual setting for the text width can be obtained with the code in
Listing 3.68.

```
\documentclass{article}
\usepackage{layouts}

\begin{document}

    textwidth in cm: \printinunitsof{cm}\prntlen{\textwidth}

\end{document}
```

**Listing 3.68**  Checking the actual text width

**Fig. 3.40** 'Sketching plane'
with $0 \leq x \leq 1$ and
$0 \leq y \leq 1$, see Listing 3.67

**Fig. 3.41** Schematic
drawing of a mechanical
member due to advanced
features, see Listing 3.69

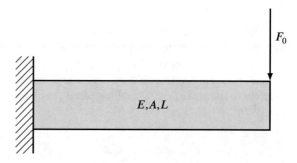

The following example shows a parameterized object from the area of mechanical engineering,[9] i.e. a structural beam which is fixed at its left-hand end and loaded by a force $F_0$ at its right-hand end, see Fig. 3.41. The horizontal length of this beam ('rectangle') is given by a variable \L and it can be positioned by a single coordinate of the lower left corner of the beam expressed by (\xs,\ys). This coordinate allows to position the object in a larger drawing. If this is not required, one may simply assign (0, 0). This example requires the TikZ library calc to calculate coordinates, the library arrows.meta to adjust the arrow head and the library decorations.pathreplacing to create a pattern which represents the fixed support. A further feature is the use of the command \clip, which allows to suppress all features of the following pattern outside the defined rectangle, see Listing 3.69.

```
1   \documentclass{article}
2   ...
3   \usepackage{tikz}
4   \usetikzlibrary{calc}
5   \usetikzlibrary{arrows.meta} % to change the size of an arrowhead
6   \usetikzlibrary{decorations.pathreplacing} % for the pattern
7   ...
```

---

[9] See [25] for details on one-dimensional structural members.

```
8    \begin{document}
9    ...
10   \begin{figure}[h!]
11   \centering
12   \begin{tikzpicture}[]
13   % Definition of Location and Size of Structural Member (Rectangle),
14   % i.e. lower left corner \xs,\ys and Length \L
15      \pgfmathparse{6.0} \pgfmathresult \let \L \pgfmathresult;
16      \pgfmathparse{0.0} \pgfmathresult \let \xs \pgfmathresult;
17      \pgfmathparse{0.0} \pgfmathresult \let \ys \pgfmathresult;
18   %
19   % Structural Member as a Rectangle
20      \draw[line width=1.0pt,fill=lightgray] (\xs, \ys) rectangle (\xs+\L,
21                                                               \ys+0.2*\L);
22   % Vertical Force
23      \draw[-{Latex[length=2mm width=1.5mm]},line width=1.0pt] (\xs+\L,
24         \ys+0.5*\L) — (\xs+\L, \ys+0.2*\L) node[pos=0.4,right]{$F_0$};
25   % Labels
26      \node[anchor = center] at (\xs + 0.5*\L,\ys+0.1*\L) {$E, \!A, \!L$};
27   % Left-Hand Fixed Suppport
28      \draw[line width=1.0pt] (\xs, \ys-0.1*\L) — (\xs, \ys+0.3*\L);
29      \clip (\xs-0.1*\L,\ys-0.1*\L) rectangle (\xs,\ys+0.3*\L);
30      \draw [preaction={decorate, draw,line width=0.5pt}, decoration={border,
31         angle=-45, amplitude=3*\L, segment length=1.75mm}](\xs,\ys+0.35*\L)
32         — (\xs,\ys-0.1*\L);
33   \end{tikzpicture}
34   \caption{Schematic drawing of a mechanical member due to advanced features}
35   \label{fig:schematic_beam_01}
36   \end{figure}
37   ...
```

**Listing 3.69** Schematic drawing of a mechanical member due to advanced features

Another possibility to automatize the generation of graphical objects is to group them in a new environment via the **scope** command. The argument of the scope environment allows to apply common transformations to all the objects within the scope environment, see Listing 3.70 and the corresponding Fig. 3.42. A user could also specify the point to rotate around by using the command **rotate around={deg:(x,y)}**, where **deg** specifies the angle and the point is given by its coordinates **(x,y)**.

```
1    \begin{figure}[h!]
2    \centering
3        \begin{tikzpicture}
4    %
5            \begin{scope}[rotate=30]
6                \draw (3,3) — (5,4);
7                \draw (3,3) circle (0.75);
8            \end{scope}
9    %
10       \end{tikzpicture}
11   \caption{Grouping of graphical objects and common transformation (rotation)}
12   \label{fig:scope_o}
13   \end{figure}
```

**Listing 3.70** Grouping of graphical objects and common transformation (rotation)

Furthermore, this group of objects can be accessed via a new command by a definition based on \newcommandx, for details see [28]. The application of this definition requires the LATEX package xargs, see Listing 3.71 and the corresponding Fig. 3.43. The new command is called \object and the number in the following square brackets, i.e. 3 in our example, specifies the number of arguments this macro will take. The arguments are then labeled as #1, #2, and #3 in the scope-environment.

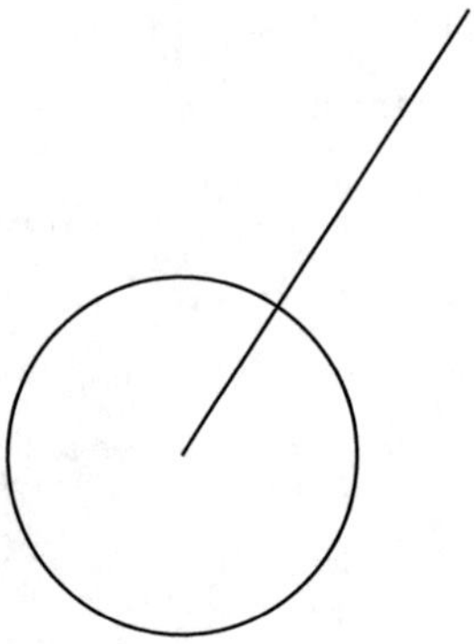

**Fig. 3.42** Grouping of
graphical objects and
common transformation
(rotation), see Listing 3.70

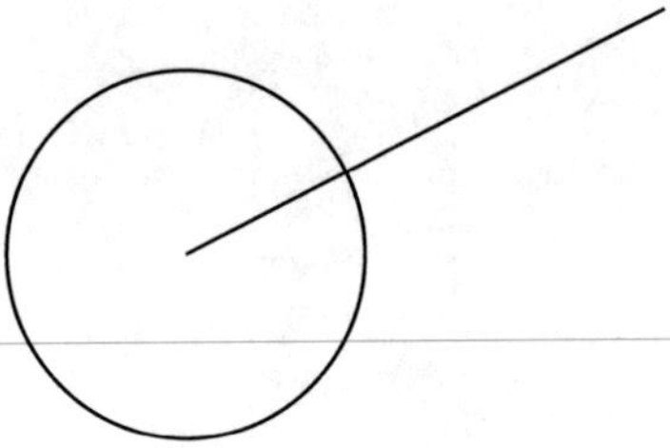

**Fig. 3.43** Grouping of
graphical objects and access
via a new command with
different arguments, see
Listing 3.71

```
1   \documentclass{article}
2   \usepackage{tikz}
3   \usepackage{xargs}
4
5   \begin{document}
6
7   \begin{figure}[h!]
8   \centering
9   \begin{tikzpicture}[]
10  %
11  \newcommandx{\object}[3][]
12          {
13              \begin{scope}
14                  \draw (#1) — (#3);
15                  \draw (#1) circle (#2);
16              \end{scope}
17          }
18  %
19  \pgfmathparse{0.75} \pgfmathresult \let \R \pgfmathresult ;
20  \coordinate (P1) at (3,3);
21  \coordinate (P2) at (5,4);
22  %
23  \object{P1}{\R}{P2};
24  \end{tikzpicture}
25  \caption{Grouping of graphical objects and access via a new ...}
26  \label{fig:scope_1}
27  \end{figure}
28
29  \end{document}
```

**Listing 3.71** Grouping of graphical objects and access via a new command with different arguments

The definition of the new command also can be done globally by defining it before
the documents starts, see Listing 3.72.

```
1   \documentclass{article}
2   \usepackage{tikz}
3   \usepackage{xargs}
4
5   \newcommandx{\object}[3][]
6           {
7           \begin{scope}
8           \draw (#1) --- (#3);
9           \draw (#1) circle (#2);
10          \end{scope}
11          }
12
13  \begin{document}
14
15  \begin{figure}[h!]
16  \centering
17  \begin{tikzpicture}[]
18  %
19  \pgfmathparse{0.75} \pgfmathresult \let \R \pgfmathresult ;
20  \coordinate (P1) at (3,3);
21  \coordinate (P2) at (5,4);
22  %
23  \object{P1}{\R}{P2};
24  \end{tikzpicture}
25  \caption{Grouping of graphical objects and access via a new ...}
26  \label{fig:scope_1}
27  \end{figure}
28
29  \end{document}
```

**Listing 3.72** Grouping of graphical objects and access via a global new command

Alternatively, the new command can be saved in an external file, for example **library.tex** and included before the document starts, see Listing 3.73.

```
1   \documentclass{article}
2   \usepackage{tikz}
3   \usepackage{xargs}
4
5   \include{library}
6
7   \begin{document}
8
9   \begin{figure}[h!]
10  \centering
11  \begin{tikzpicture}[]
12  %
13  \pgfmathparse{0.75} \pgfmathresult \let \R \pgfmathresult ;
14  \coordinate (P1) at (3,3);
15  \coordinate (P2) at (5,4);
16  %
17  \object{P1}{\R}{P2};
18  \end{tikzpicture}
19  \caption{Grouping of graphical objects and access via a new ...}
20  \label{fig:scope_1}
21  \end{figure}
22
23  \end{document}
```

**Listing 3.73** Grouping of graphical objects and access via a global new command from an external file

## 3.3   Tables

### 3.3.1   *Import from External Files*

Let us consider in the following a data table with the file name data_table_
pgftable_1.txt, which is located in a subdirectory ('tables') of the actual working
directory. This ASCII file[10] contains three columns which represent 11 data points
$(x_1, y_1, y_2)$.

```
x_1      y_1          y_2
0        1.0012       2.0313
1        1.4003       2.4403
2        1.4338       2.4934
3        1.4819       2.4011
4        1.8215       2.8319
5        2.           2.4124
6        2.3002       2.3103
7        2.5916       2.5817
8        2.7508       2.
9        2.9110       2.9911
10       3.0738       4.0239
```

The following Listing 3.74 shows a way to load this data file and to automatically
create a simple LATEXtable, see Table 3.14 for the compiled result. This requires
the incorporation of the LATEX package pgfplotstable. The underlying table definition is embedded in the \pgfplotstabletypeset[...] environment and the command
columns/.../.style={...} allows the definition of each column.

```
1   \documentclass{article}
2   \usepackage{pgfplotstable}
3   ...
4   \begin{document}
5   ...
6   \begin{table}[h!]
7       \begin{center}
8       \caption{Autogenerated table external text file}\label{tab:table_ex_1}
9       \ pgfplotstabletypeset[
10      columns/x_1/.style={int detect,column type=c,column name=$x_1$},
11      columns/y_1/.style={sci,sci zerofill,sci sep align,precision=2,
12                          sci 10e,column name=$y_1$,},
13      columns/y_2/.style={fixed,fixed zerofill,precision=3,column name=$y_2$} ]
14      {tables/data_table_pgftable_1.txt}
15      \end{center}
16  \end{table}
```

**Listing 3.74**  Autogenerated table from external text file

The number format definitions in Listing 3.74 are mostly included in Table 3.2.

   Some ASCII tables may come without any column names and the name 'x_1', for
example, in line 10 of Listing 3.74 could not be addressed as columns/x_1/. Then,
the command display columns allows to address the column number (starting from
0). Listing 3.74 could be modified as follows:

---

[10] Lines starting with % or # are comments and are not imported/read.

**Table 3.14** Autogenerated table from external text file, see Listing 3.74

| $x_1$ | $y_1$ | $y_2$ |
|---|---|---|
| 0 | $1.00 \cdot 10^0$ | 2.031 |
| 1 | $1.40 \cdot 10^0$ | 2.440 |
| 2 | $1.43 \cdot 10^0$ | 2.493 |
| 3 | $1.48 \cdot 10^0$ | 2.401 |
| 4 | $1.82 \cdot 10^0$ | 2.832 |
| 5 | $2.00 \cdot 10^0$ | 2.412 |
| 6 | $2.30 \cdot 10^0$ | 2.310 |
| 7 | $2.59 \cdot 10^0$ | 2.582 |
| 8 | $2.75 \cdot 10^0$ | 2.000 |
| 9 | $2.91 \cdot 10^0$ | 2.991 |
| 10 | $3.07 \cdot 10^0$ | 4.024 |

```
9    ...
10   display columns/0/.style={...
11   display columns/1/.style={...
12   ...
```

**Listing 3.75** Addressing column numbers of the source file

In the case of long tables, it might be requested to not plot all rows of the original ASCII table. The command skip rows between index={0}{4} would skip, for example, the first five rows. This command could be included in line 9 of Listing 3.74. It should be noted here that this command can be applied multiple times to refine the selection, e.g. selecting specific blocks from the original table. External programs may generate ASCII tables which are based on different types of separators. The command col sep=... with the possible arguments space, tab, comma, semicolon, colon, braces, &, and ampersand allows to address this in the pgfplotstabletypeset (e.g. in line 9 of Listing 3.74) environment.

The layout of Table 3.14 can be further improved by adding horizontal lines as shown in Listing 3.76 and Table 3.15. It should be noted here that the commands \toprule, \midrule and \bottomrule require the package booktabs.

```
1    \documentclass{article}
2    \usepackage{pgfplotstable}
3    \usepackage{booktabs}
4    ...
5    \begin{document}
6    ...
7    \begin{table}[h!]
8    \begin{center}
9    \caption{Autogenerated table with horizontal lines}
10   \label{tab:pgfsplotstable_02}
11       \pgfplotstabletypeset[
12       columns/x_1/.style={int detect,column type=c,column name=$x_1$ in mm},
13       columns/y_1/.style={sci,sci zerofill,sci sep align,precision=2,sci 10e,
14                          column name=$y_1$ in mm},
15       columns/y_2/.style={fixed,fixed zerofill,precision=3,column name=$y_2$
16                          in mm},
17       every head row/.style={before row=\toprule,after row=\midrule},
18       every last row/.style={after row=\bottomrule} ]
19       {tables/data_table_pgftable_1.txt}
20   \end{center}
21   \end{table}
```

**Table 3.15**  Autogenerated table with horizontal lines, see Listing 3.76

| $x_1$ in mm | $y_1$ in mm | $y_2$ in mm |
|---|---|---|
| 0 | $1.00{\cdot}10^0$ | 2.031 |
| 1 | $1.40{\cdot}10^0$ | 2.440 |
| 2 | $1.43{\cdot}10^0$ | 2.493 |
| 3 | $1.48{\cdot}10^0$ | 2.401 |
| 4 | $1.82{\cdot}10^0$ | 2.832 |
| 5 | $2.00{\cdot}10^0$ | 2.412 |
| 6 | $2.30{\cdot}10^0$ | 2.310 |
| 7 | $2.59{\cdot}10^0$ | 2.582 |
| 8 | $2.75{\cdot}10^0$ | 2.000 |
| 9 | $2.91{\cdot}10^0$ | 2.991 |
| 10 | $3.07{\cdot}10^0$ | 4.024 |

**Listing 3.76**  Autogenerated table with horizontal lines

In case that the package booktabs is not available, one may simply replace the commands \toprule, \midrule and \bottomrule by \noalign{\smallskip} \hline \noalign{\smallskip}. It is obvious that the pgfplotstable package is not able to introduce all formatting options as in the case of writing the entire LATEXcode. Depending on the task, one may request to export the table as LATEXcode and then introduce further formatting in this code. This can be done, for example, as shown in Listing 3.76 by including the command outfile=....tex in the \pgfplotstabletypeset[...] environment. The updated code is shown in Listing 3.77.

```
1    \documentclass{article}
2    \usepackage{pgfplotstable}
3    \usepackage{booktabs}
4    ...
5    \begin{document}
6    ...
7    \begin{table}[h!]
8    \begin{center}
9    \caption{Autogenerated table with horizontal lines}
10   \label{tab:pgfsplotstable_02}
11       \ pgfplotstabletypeset[
12       columns/x_1/.style={int detect,column type=c,column name=$x_1$ in mm},
13       columns/y_1/.style={sci,sci zerofill,sci sep align,precision=2,sci 10e,
14                       column name=$y_1$ in mm},
15       columns/y_2/.style={fixed,fixed zerofill,precision=3,column name=$y_2$
16                       in mm},
17       every head row/.style={before row=\toprule,after row=\midrule},
18       every last row/.style={after row=\bottomrule},outfile=Latex_code.tex]
19   {tables/data_table_pgftable_1.txt}
20   \end{center}
21   \end{table}
```

**Listing 3.77**  Export of autogenerated table to LATEX format

The created LᴬTᴇX code in file Latex_code.tex looks as follows:

```
\begin {tabular}{cr<{\pgfplotstableresetcolortbloverhangright }@{}|
<{\pgfplotstableresetcolortbloverhangleft }c}%
\toprule $x_1$ in mm&\multicolumn {2}{c}{$y_1$ in mm}&$y_2$ in mm\\ midrule %
\pgfutilensuremath {0}&$1.00$&$\cdot 10^{0}$&\pgfutilensuremath {2.031}\\%
\pgfutilensuremath {1}&$1.40$&$\cdot 10^{0}$&\pgfutilensuremath {2.440}\\%
\pgfutilensuremath {2}&$1.43$&$\cdot 10^{0}$&\pgfutilensuremath {2.493}\\%
\pgfutilensuremath {3}&$1.48$&$\cdot 10^{0}$&\pgfutilensuremath {2.401}\\%
\pgfutilensuremath {4}&$1.82$&$\cdot 10^{0}$&\pgfutilensuremath {2.832}\\%
\pgfutilensuremath {5}&$2.00$&$\cdot 10^{0}$&\pgfutilensuremath {2.412}\\%
\pgfutilensuremath {6}&$2.30$&$\cdot 10^{0}$&\pgfutilensuremath {2.310}\\%
\pgfutilensuremath {7}&$2.59$&$\cdot 10^{0}$&\pgfutilensuremath {2.582}\\%
\pgfutilensuremath {8}&$2.75$&$\cdot 10^{0}$&\pgfutilensuremath {2.000}\\%
\pgfutilensuremath {9}&$2.91$&$\cdot 10^{0}$&\pgfutilensuremath {2.991}\\%
\pgfutilensuremath {10}&$3.07$&$\cdot 10^{0}$&\pgfutilensuremath {4.024}\\
\bottomrule %
\end {tabular}%
```

This code can be directly compiled but requires the pgfplotstable and booktabs package. To have a more plain LᴬTᴇXcode, one may delete r<{\pgfplotstableresetcolortbloverhangright } and <{\pgfplotstableresetcolortbloverhangleft } and replace \pgfutilensuremath by \ensuremath. In addition, the commands related to the vertical rules must be replaced by \noalign{\smallskip}\hline**\noalign{\smallskip}**. The following code can be easily compiled to obtain the same result as shown in Table 3.15.

```
\begin {tabular}{cr@{}|c}%
\noalign{\smallskip}\hline\noalign{\smallskip} $x_1$ in mm&
\multicolumn {2}{c}{$y_1$ in mm}&$y_2$ in mm\\
\noalign{\smallskip}\hline\noalign{\smallskip} %
\ensuremath {0}&$1.00$&$\cdot 10^{0}$&\ensuremath {2.031}\\%
\ensuremath {1}&$1.40$&$\cdot 10^{0}$&\ensuremath {2.440}\\%
\ensuremath {2}&$1.43$&$\cdot 10^{0}$&\ensuremath {2.493}\\%
\ensuremath {3}&$1.48$&$\cdot 10^{0}$&\ensuremath {2.401}\\%
\ensuremath {4}&$1.82$&$\cdot 10^{0}$&\ensuremath {2.832}\\%
\ensuremath {5}&$2.00$&$\cdot 10^{0}$&\ensuremath {2.412}\\%
\ensuremath {6}&$2.30$&$\cdot 10^{0}$&\ensuremath {2.310}\\%
\ensuremath {7}&$2.59$&$\cdot 10^{0}$&\ensuremath {2.582}\\%
\ensuremath {8}&$2.75$&$\cdot 10^{0}$&\ensuremath {2.000}\\%
\ensuremath {9}&$2.91$&$\cdot 10^{0}$&\ensuremath {2.991}\\%
\ensuremath {10}&$3.07$&$\cdot 10^{0}$&\ensuremath {4.024}\\
\noalign{\smallskip}\hline\noalign{\smallskip} %
\end {tabular}%
```

Let us introduce in the following one more formatting option, which may be quite useful. Listing 3.78 and the corresponding Table 3.16 show a way to introduce a second headrow which, in this example, contains the units of the measurements.

```
1  \documentclass{article}
2  \usepackage{pgfplotstable}
3  \usepackage{booktabs}
4  ...
5  \begin{document}
6  ...
7  \begin{table}[h!]
8  \begin{center}
9  \caption{Automated generation of a second headrow}
10 \label{tab:table_ex_04}
11     \pgfplotstabletypeset[multicolumn names,
12     columns/x_1/.style={int detect,,column type=c,column name=$x_1$},
13     columns/y_1/.style={sci,sci zerofill,precision=2,sci 10e,
14                         column name=$y_1$},
```

**Table 3.16**  Automated generation of a second headrow, see Listing 3.78

| $x_1$ in mm | $y_1$ in mm | $y_2$ in mm |
|---|---|---|
| 0 | $1.00 \cdot 10^0$ | 2.031 |
| 1 | $1.40 \cdot 10^0$ | 2.440 |
| 2 | $1.43 \cdot 10^0$ | 2.493 |
| 3 | $1.48 \cdot 10^0$ | 2.401 |
| 4 | $1.82 \cdot 10^0$ | 2.832 |
| 5 | $2.00 \cdot 10^0$ | 2.412 |
| 6 | $2.30 \cdot 10^0$ | 2.310 |
| 7 | $2.59 \cdot 10^0$ | 2.582 |
| 8 | $2.75 \cdot 10^0$ | 2.000 |
| 9 | $2.91 \cdot 10^0$ | 2.991 |
| 10 | $3.07 \cdot 10^0$ | 4.024 |

```
15        columns/y_2/.style={fixed,fixed zerofill,precision=3,column name=$y_2$},
16        %
17     every head row/.style={before row={\toprule},\ after row={in mm & in mm &
18                            in mm \\\midrule}},
19     every last row/.style={after row=\bottomrule} %
20   {tables/data_table_pgftable_1.txt}
21 \end{center}
22 \end{table}
```

**Listing 3.78**  Automated generation of a second headrow

The final example in Listing 3.79 and Table 3.17 shows how to automatically perform mathematical operations on entire columns. In our case, the first column of the original ASCII file is multiplied by four and introduced as the new axis $x_2$.

```
1  \documentclass{article}
2  \usepackage{pgfplotstable}
3  \usepackage{booktabs}
4  ...
5  \begin{document}
6  ...
7  \begin{table}[h!]
8  \pgfplotstableset{create on use/new/.style={create col/expr={\thisrow{x_1}*4}}}
9  \begin{center}
10 \caption{Mathematical processing during import}
11 \label{tab:table_ex_05}
12    \ pgfplotstabletypeset[
13      columns={x_1, new, y_1, y_2},
14      columns/x_1/.style={int detect,column type=c,column name=$x_1$ in mm},
15      columns/new/.style={int detect,column type=c,column name=$x_2$ in mm},
16      columns/y_1/.style={sci,sci zerofill,sci sep align,precision=2,sci 10e,
17                          column name=$y_1$ in mm},
18      columns/y_2/.style={fixed,fixed zerofill,precision=3,column name=$y_2$
19                          in mm},
20    every head row/.style={before row=\toprule,after row=\midrule},
21    every last row/.style={after row=\bottomrule} ]
22    {tables/data_table_pgftable_1.txt}
23 \end{center}
24 \end{table}
```

**Listing 3.79**  Mathematical processing during import

**Table 3.17** Mathematical processing during import, see Listing 3.79

| $x_1$ in mm | $x_2$ in mm | $y_1$ in mm | $y_2$ in mm |
|---|---|---|---|
| 0 | 0 | $1.00 \cdot 10^0$ | 2.031 |
| 1 | 4 | $1.40 \cdot 10^0$ | 2.440 |
| 2 | 8 | $1.43 \cdot 10^0$ | 2.493 |
| 3 | 12 | $1.48 \cdot 10^0$ | 2.401 |
| 4 | 16 | $1.82 \cdot 10^0$ | 2.832 |
| 5 | 20 | $2.00 \cdot 10^0$ | 2.412 |
| 6 | 24 | $2.30 \cdot 10^0$ | 2.310 |
| 7 | 28 | $2.59 \cdot 10^0$ | 2.582 |
| 8 | 32 | $2.75 \cdot 10^0$ | 2.000 |
| 9 | 36 | $2.91 \cdot 10^0$ | 2.991 |
| 10 | 40 | $3.07 \cdot 10^0$ | 4.024 |

### 3.3.2 Spreadsheet Analysis

The following descriptions are based on the LaTeXpackage spreadtab, which allows, to some degree, classical spreadsheet analyses [42]. Additional options are based on the packages fp, SageTeX, or tabularcalc. Using the spreadtab package, a new environment \begin{spreadtab}...\end{spreadtab} can be used within a tabular (or any other LaTeXtable) environment. Each cell has now its unique index: columns are indicated by alphabetical letters a to z and rows by Roman numbers starting with 1. The following Listing 3.80 and Table 3.18 show an example where the first three rows are summed up in the third row and the first two columns are summed up in row four. It can be seen that, for example, the upper left cell is addressed by its index 'a1'.

```
1  \begin{table}[h!]
2      \begin{center}
3          \caption{Simple calculation in a table}
4          \label{tab:spreadtab_01}
5              \begin{spreadtab}{{tabular}{rr|r}}
6                  19 &  9  & a1+b1   \\
7                  22 & 16  & a2+b2   \\
8                   4 & 11  & a3+b3   \\ \hline
9                  a1+a2+a3 & b1+b2+b3 &
10             \end{spreadtab}
11     \end{center}
12 \end{table}
```

**Listing 3.80** Simple calculation in a table

The following Listing 3.81 results in the same Table 3.18 but introduces relative referencing (based on horizontal and vertical shift from the actual cell indicated in square brackets [...,...]). In addition, assigning (tag) and referencing (cell) to user-defined cell names as well as the summation function (sum) are applied.

```
1  \begin{table}[h!]
2      \begin{center}
3          \caption{Simple calculation in a table}
4          \label{tab:spreadtab_01}
```

**Table 3.18**  Simple calculation in a table, see Listing 3.80

| 19 | 9 | 28 |
|---:|---:|---|
| 22 | 16 | 38 |
| 4 | 11 | 15 |
| 45 | 36 | |

**Table 3.19**  Text in cells and advanced global use of assignments, see Listing 3.82

| $x$ | $y$ | $z$ |
|---:|---:|---|
| 19 | 9 | 5 |
| 22 | 16 | 8 |
| 4 | 11 | 12 |
| | Sum $z$: 25 | |

```
5      \begin{spreadtab}{{tabular}{rr|r}}
6            19 &  9 &  [-2,0]+[-1,0] \\
7            22 & 16 & sum(a2:b2) \\
8             4 tag(ksk) & 11 &  cell(ksk)+b3 \\ \hline
9          sum(a1:[0,−1] ) & b1+b2+b3 &
10         \end{spreadtab}
11      \end{center}
12   \end{table}
```

**Listing 3.81**  Calculation in a table: relative referencing, assigning cell names, and sum function

The following Listing 3.82 introduces text cells. For pure text or comment cells, the special character @ is used to indicate a non-mathematical content. For mixed cells, i.e. with text and mathematical content, the mathematical content must be enclosed in :={...}. Another feature is the global definition (textsf\STmakegtag{...}) of variables, which can be even used (\STtag{...}) outside the table environment. The corresponding output is shown in Table 3.19.

```
1    \begin{table}[h!]
2    \setlength{\tabcolsep}{6pt}
3          \begin{center}
4          \caption{Text in cells and advanced global use of assignments}
5          \label{tab:_spreadtab_03}
6                \begin{spreadtab}{{tabular}{rr|r}}
7                   @ $x$ &  @  $y$ &  @ $z$ \\ \hline
8                   19 &  9 & 5 \\
9                   22 & 16 & 8 \\
10                  4 & 11 & 12 \\
11                  \multicolumn{3}{r}{Sum $z$:  :={sum(c2:c4)tag(erg) }}\\
12               \end{spreadtab}
13          \STmakegtag{erg}
14          \end{center}
15   \end{table}
16   ...
17   The sum of \STtag{erg} is  ...
```

**Listing 3.82**  Text in cells and advanced global use of assignments

The mathematical calculations presented in Listings 3.80–3.82 were restricted to simple summations of numbers. However, the spreadtab environment allows to apply a variety of mathematical operations which are collected in Listing 3.83

**Table 3.20** Available mathematical operations (arguments of the trigonometric functions should be provided in radian), see Listing 3.83

| $x$-value | 1 | 4 | 3 |
|---|---|---|---|
| $y$-value | 2 | 3 | 7 |
| $f(x) = 2x$ | 2 | 8 | 6 |
| $f(x, y) = x \times y$ | 2 | 12 | 21 |
| $f(x, y) = x + y$ | 3 | 7 | 10 |
| $f(x, y) = x - y$ | -1 | 1 | -4 |
| $f(x, y) = x/y$ | 0,5 | 1,333 | 0,429 |
| $f(x, y) = x^y$ | 1 | 64 | 2187 |
| $f(x) = \ln(x)$ | 0 | 1,386 | 1,099 |
| $f(x) = \sin(x)$ | 0,841 | -0,757 | 0,141 |
| $f(x) = \sqrt{x}$ | 1 | 2 | 1,732 |

and Table 3.20. In addition, the listing shows how to set the rounding of numbers based on \STautoround{...} and how to set the decimal separator to a comma with \STsetdecimalsep{...}.

```
\begin{table}[h!]
    \begin{center}
        \caption{Available mathematical operations in a table}
        \label{tab:_spreadtab_04}
            \STautoround{3} %\STautoround*{3} would fill with 0
            \STsetdecimalsep{{,}}
            \begin{spreadtab}{{tabular}{l|lccc|}}
            \hline
                @ $x$-value & 1 & 4 & 3 \\
                @ $y$-value & 2 & 3 & 7 \\ \hline
                @ $f(x)=2x$ & 2 *[0,-2] & 2*[0,-2] & 2*[0,-2]\\
                @ $f(x,y)=x\times y$ & [0,-3] *[0,-2] & [0,-3]*[0,-2] &
                                                    [0,-3]*[0,-2]\\
                @ $f(x,y)=x + y$ & b1 + b2 & c1+c2 & d1+d2\\
                @ $f(x,y)=x - y$ & b1 - b2 & c1-c2 & d1-d2\\
                @ $f(x,y)=x / y$ & b1 / b2 & c1/c2 & d1/d2\\
                @ $f(x,y)=x^y$ & b1^b2 & c1^c2 & d1^d2\\
                @ $f(x)=\ln(x)$ & ln(b1) & ln(c1) & ln(d1)\\
                @ $f(x)=\sin(x)$ & sin(b1) & sin(c1) & sin(d1)\\
                @ $f(x)=\sqrt{x}$ & b1^(0.5) & c1^(0.5) & d1^(0.5)\\
            \hline
        \end{spreadtab}
    \end{center}
\end{table}
```

**Listing 3.83**  Available mathematical operations in a table

Looking at line 11 in Listing 3.83, one can see that each cell contains the same equation. Thus, a copy function could reduce the work to write all the cell's contents. This can be achieved with the command \STcopy{>x,vy}{formula} where $x$ and $y$ are positive numbers that represent horizontal (> can be understood as to the right-hand side) and vertical (v can be understood as downwards) offsets relative to the cell where the command is used. Thus, the updated and shorter code for line 11 could look as indicated in Listing 3.85.

**Table 3.21**  Statistical evaluation of a data set, see Listing 3.85

| $x_i$ | 172 | 169 | 177 | 178 | 178 | 179 | 180 | 191 | 202 |
|---|---|---|---|---|---|---|---|---|---|
| | | | | mean value $\overline{x}$: 180.7 | | | | | |
| $(x_i - \overline{x})$ | -8.7 | -11.7 | -3.7 | -2.7 | -2.7 | -1.7 | -0.7 | 10.3 | 21.3 |
| $(x_i - \overline{x})^2$ | 75.7 | 136.9 | 13.7 | 7.3 | 7.3 | 2.9 | 0.5 | 106.1 | 453.7 |

$$\sum_i^n (x_i - \overline{x})^2: 804.1$$

statistics            variance: $v = \frac{1}{n} \sum_i^n (x_i - \overline{x})^2$: 89.3

standard deviation: $s = \sqrt{v} = (\frac{1}{n} \sum_i^n (x_i - \overline{x})^2)^{0.5}$: 9.5

```
10   ...
11        @ $f(x)=2x$      & \STcopy{>}{2*[0,-2]} &  &  \\
12   ...
```

**Listing 3.84**  Automated copying of a formula to other cells

The code could have been also written in the following way: \STcopy{>}{2*[!0,!-2]}.
The exclamation point (!) would cause the coordinate in the formula to not be modified, and thus all cells would be calculated the same, i.e. 2*b2.

Let us now look at the statistical evaluation of a data set. Assuming that nine measured values of a certain quantity ($x_i$) are given, the mean value and the standard deviation are to be calculated. Listing 3.85 and Table 3.21 show a possible solution.

```
1   \begin{table}[h!]
2       \begin{center}
3       \caption{Statistical evaluation of a data set}
4       \label{tab:_spreadtab_05}
5           \STautoround{1}
6           \begin{spreadtab}{{tabular}{ crrrrrrrr }} \hline
7           @ $x_i$  & 172tag(first) & 169 & 177 & 178 & 178 &, 179 & 180 &
8           191 & 202tag(last) \\ \hline
9           & \multicolumn{9}{c}{mean value $\overline{x}$: :={sum(cell
10          (first):cell(last))/(col(last)-col(first)+1)tag(ave)}}
11          \rule[-2.5mm]{0mm}{6.5mm}\\ \hline
12          @ $(x_i-\overline{x})$  & \STcopy{>}{([0,-2]-cell(ave))} & &
13          & & & & & \\
14          @ $(x_i-\overline{x})^2$  & \STcopy{>}{abs([0,-3]-cell(ave))
15          ^2} & & & & & & & &\\ \hline
16          & \multicolumn{9}{c}{$\sum \limits_i^n(x_i-\overline{x})^2$:
17          :={sum(b4:j4)}} \rule[-2.5mm]{0mm}{6.5mm}\\
18          @ statistics & \multicolumn{9}{c}{variance: $v=\frac{1}{n}\sum
19          \limits_i^n(x_i-\overline{x})^2$: :={sum(b4:j4)/(col(last)-col
20          (first)+1)}} \rule[-2.5mm]{0mm}{6.5mm}\\
21          & \multicolumn{9}{c}{standard deviation: $s=\sqrt{v}=(\frac{1}{n}
22          \sum \limits_i^n(x_i-\overline{x})^2)^{0.5}$: :={sum(b4:j4)/
23          (col(last)-col(first)+1))^(0.5)}} \rule[-2.5mm]{0mm}{6.5mm}\\
24          \hline
25      \end{spreadtab}
26      \end{center}
27  \end{table}
```

**Listing 3.85**  Statistical evaluation of a data set

**Table 3.22** Statistical evaluation of a data set: example 2, see Listing 3.86

| $x_i$ | 1 | 2 | 3 | 4 | 5 | 6 |
|---|---|---|---|---|---|---|
| $n_i$ | 4 | 5 | 7 | 6 | 2 | 1 |

$$\overline{x} = \frac{1}{n} \times \sum_{i=1}^{n} (x_i \times n_i): 3$$

| $(x_i - \overline{x})$ | -2 | -1 | 0 | 1 | 2 | 3 |
|---|---|---|---|---|---|---|
| $(x_i - \overline{x})^2 \times n_i$ | 16 | 5 | 0 | 6 | 8 | 9 |

$$\sum_{i=1}^{n} (x_i - \overline{x})^2 \times n_i: 44$$

statistics
$$v = \frac{1}{n} \times \sum_{i=1}^{n} (x_i - \overline{x})^2 \times n_i: 1.76$$

$$s = \sqrt{v} = (\frac{1}{n} \times \sum_{i=1}^{n} (x_i - \overline{x})^2 \times n_i)^{0.5}: 1.33$$

Some of the results presented in Table 3.21 are only intermediary results. The commands \SThiderow and \SThidecol hide a row or column, when placed in a cell.

The final example is a slight modification of the previous one. Let us assume that six classes $(x_i)$ are now given and a certain number $(n_i)$ is recorded for each class. This could be, for example, the grading scale and the numbers of students which achieved a certain grade. Listing 3.86 and Table 3.22 offer a possible solution.

```
1   \begin{table}[h!]
2       \begin{center}
3       \caption{Statistical evaluation of a data set: example 2}
4       \label{tab:_spreadtab_06}
5           \STautoround{2}
6           \begin{spreadtab}{{tabular}{crrrrrr}} \hline
7               @ $x_i$  & 1tag(first)& 2 & 3 & 4 & 5 & 6tag(last)\\
8               @ $n_i$  & 4tag(sta) & 5 & 7 & 6 & 2 & 1tag(end)\\
9               \hline
10              & \multicolumn{6}{c}{$\overline{x}=\frac{1}{n}\times \sum
11              \limits_{i\,=\,1}^n (x_i\times n_i)$: :={sumprod(cell(first):
12              cell(last);cell(sta):cell(end))/sum(cell(sta):cell(end))tag
13              (ave)}} \rule[-2.5mm]{0mm}{6.5mm}\\ \hline
14              @ $(x_i-\overline{x})$  & \STcopy{>}{([0,-3]-cell(ave))} & & &
15              & & \\
16              @ $(x_i-\overline{x})^2\times n_i$  & \STcopy{>}{abs([0,-4]-
17              cell(ave))^2*[0,-3]} & & & & &\\
18              \hline
19              & \multicolumn{6}{c}{$\sum \limits_{i\,=\,1}^n(x_i-\overline{x}
20              )^2\times n_i$: :={sum(b5:g5)tag(sump)}}
21              \rule[-2.5mm]{0mm}{6.5mm}\\
22              @ statistics & \multicolumn{6}{c}{$v=\frac{1}{n}\sum \limits_{i
23              \,=\,1}^n(x_i-\overline{x})^2\times n_i$: :={cell(sump)/(sum(
24              cell(sta):cell(end))tag(varia)))} \rule[-2.5mm]{0mm}{6.5mm}\\
25              & \multicolumn{6}{c}{ $s=\sqrt{v}=(\frac{1}{n}\sum \limits_{i
26              \,=\,1}^n(x_i-\overline{x})^2\times n_i)^{0.5}$: :={cell(varia
27              )^(0.5)}} \rule[-2.5mm]{0mm}{6.5mm}\\ \hline
28          \end{spreadtab}
29      \end{center}
30  \end{table}
```

**Listing 3.86** Statistical evaluation of a data set: example 2

# Chapter 4
# Presentations

**Abstract** This chapter introduces the beamer document class which allows to generate slide shows. The second part of this chapter is devoted to the generation of posters.

## 4.1 Slide Shows

Different specialized document classes, such as beamer [6], prosper [15], powerdot [1], or scrartcl [21] are available to produce slide shows. However, in this chapter we will focus on the BEAMER document class, which was originally developed by Till Tantau and used for his PhD defense presentation (Dr. rer.-nat.) in February 2003 at the TU Berlin. This class comes with more than 240 pages of documentation, see [40] for details. The document class must be, as all other classes, initiated in the preamble of the document, see Listing 4.1.

```
1  \documentclass[...]{beamer}
2  ...
3  \begin{document}
4  ...
```

**Listing 4.1** Initialization of the BEAMER class in the preamble

Possible beamer class options are summarized in Table 4.1.

### 4.1.1 Defining and Structuring a Frame (Slide)

#### 4.1.1.1 Title Page

The creation of a default title page requires the definition of some elements,[1] i.e., title, subtitle, authors, institutions and date, in the preamble of the document, see Listing 4.2 and Fig. 4.1 for the graphical output. All these commands allow addition-

---

[1] If any of these definitions are missing, they are not shown on the title page.

**Table 4.1** Some of the BEAMER class options

| Option | Comment |
| --- | --- |
| compress | Makes all navigation bars as small as possible |
| draft | Graphics, headlines, footlines are replaces by gray rectangles to speed up compiling |
| handout | For PDF handouts whereas each slide is one time shown |
| mathserif/mathsans | Uses fonts for the text with serif for maths (default is to use sans-serif fonts as for the text) |
| sans/serif | Uses fonts for the text with or without serif (default is to use sans-serif fonts) |
| slidestop | Puts frame titles on top left corner (default is slidescentered) |
| trans | To create a PDF transparencies version. As default, all overlay specifications are suppressed |
| xpt | Default font size is 11 pt but may take the following values: 8 pt, 9 pt, 10 pt, 11 pt, 12 pt, 14 pt, 17 pt, and 20 pt |

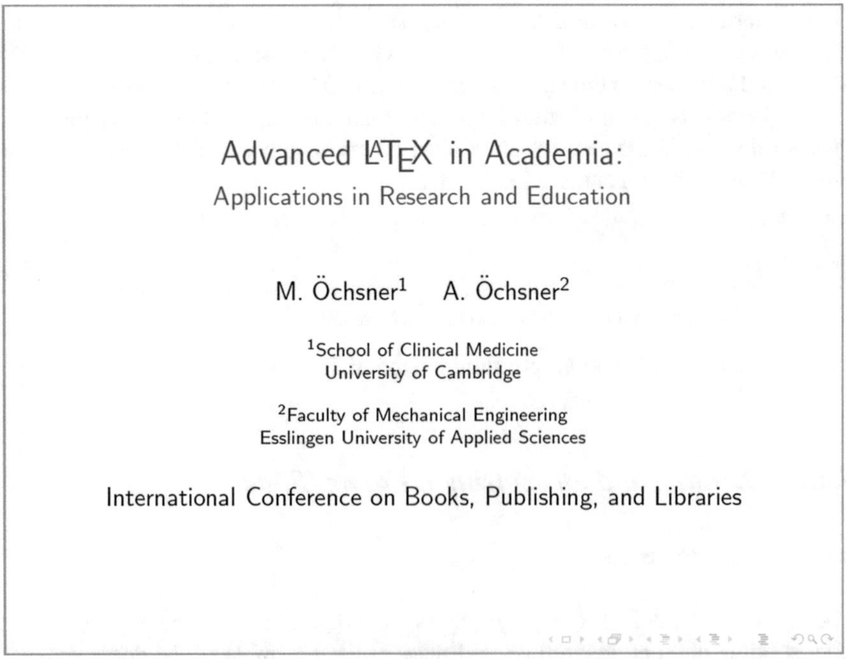

## Advanced LATEX in Academia:
### Applications in Research and Education

M. Öchsner[1]    A. Öchsner[2]

[1]School of Clinical Medicine
University of Cambridge

[2]Faculty of Mechanical Engineering
Esslingen University of Applied Sciences

**International Conference on Books, Publishing, and Libraries**

**Fig. 4.1** A title page with default settings, see Listing 4.2

ally indicating a shorter form as an optional parameter in square brackets. In case that the actual date is required, one may use the command \date{\today}.

```
1   \documentclass{beamer}
2
3   \title[Advanced \LaTeX] {Advanced \LaTeX\, in Academia:}
4   \subtitle{Applications in Research and Education}
5   \author[\"{O}chsner, Marco] {M.~\"{O}chsner\inst{1} \and
6                               A.~\"{O}chsner\inst{2}}
7   \institute[short name]
8            {\inst{1}%
9            School of Clinical Medicine\newline
10           University of Cambridge
11           \and
12           \inst{2}%
13           Faculty of Mechanical Engineering\newline
14           Esslingen University of Applied Sciences
15           }
16  \date[ICBPL]{International Conference on Books, Publishing, and Libraries}
17
18
19  \begin{document}
20
21      \begin{frame}
22          \maketitle
23      \end{frame}
24
25  \end{document}
```

**Listing 4.2** A title page with default settings, see Fig. 4.1

There is also the optional argument 'plain' for the frame environment. This produces a title page that fills the whole frame: \begin{frame}[plain]\maketitle \end{frame}.

### 4.1.1.2 Regular Slide

Content that should be displayed on your slide is commonly enclosed in the \begin{frame}[options]{frame title}...\end{frame} environment, see Listing 4.3

```
1   \documentclass{beamer}
2
3   \begin{document}
4
5       \begin{frame}{A Simple Frame (Slide)}
6
7           This slide is your first step in preparing a presentation.
8
9       \end{frame}
10
11  \end{document}
```

**Listing 4.3** Definition of a regular slide (frame)

The frame title can alternatively defined by the command \frametitle{...}. In addition, there is an optional command to define a subtitle, see Listing 4.4 and Fig. 4.2.

The possible options for the frame command are summarized in Table 4.2.

```
1   \documentclass{beamer}
2
3   \begin{document}
4
5       \begin{frame}
```

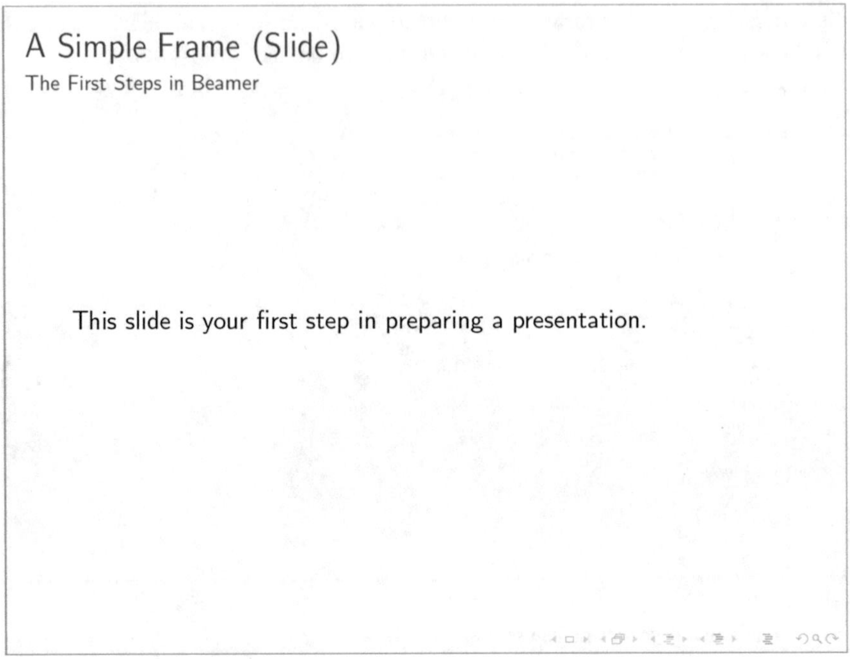

**Fig. 4.2** Definition of a regular slide: title (alternative way) and subtitle, see Listing 4.4

**Table 4.2** Some of the options for the frame environment, see [40]

| Command | Comment |
|---|---|
| allowframebreaks | The frame will be automatically distributed over several frames if the text does not fit on a single slide. This option does not allow any overlays (see Sect. 4.1.2) and is strongly recommended only for long bibliographies (see Sect. 4.1.1.3) |
| b, c, t | Content is vertically aligned at the bottom (b), center (c), or top (t) |
| noframenumbering | The framenumber counter is not increased |
| label=⟨name⟩ | The content is stored under ⟨name⟩ for later resumption using the command \againframe |
| plain | Headlines, footlines, and sidebars are suppressed |
| shrink= ⟨percentage⟩ | The text of the frame will be shrunk at least by this percentage |
| squeeze | All vertical spaces in the text are squeezed together |

```
6    \frametitle{A Simple Frame (Slide)}
7    \framesubtitle{The First Steps in Beamer}
8
9    This slide is your first step in preparing a presentation.
10
11   \end{frame}
12
13 \end{document}
```

**Listing 4.4** Definition of a regular slide: title (alternative way) and subtitle, see Fig. 4.2

As in the case of books or theses, a BEAMER presentation can be structured by sections and subsections. The corresponding commands are \section[Short Name]{Name}, \subsection[Short Name]{Name}, and \subsubsection[Short Name]{Name}. These commands are used to form the table of contents (see Sect. 4.1.1.3) based on the {Name} definition. The [Short Name] is used in conjunction with the \useoutertheme{sidebar} and \useoutertheme{infolines} definition (see Sect. 4.1.6 for details). A separator slide can be introduced by the command \begin{frame} \sectionpage \end{frame}, see Listing 4.5 and Fig. 4.3.

```
1  \documentclass{beamer}
2
3  \begin{document}
4
5  \section[Intro]{Introduction}
6      \begin{frame}
7          \sectionpage
8      \end{frame}
9      \begin{frame}{A Simple Frame (Slide)}
10         This slide is defined within a certain section.
11     \end{frame}
12
13 \end{document}
```

**Listing 4.5** Definition of a separator slide, see Fig. 4.3

To structure a slide, the block and column structures can be used. A block structure is initiated by the command \begin{block}{title}...\end{block} and divides a slide horizontally into a headed section. There is also the modification \begin{alertblock}... and \begin{exampleblock}, which result in a different layout. The column structure is initiated by the command \begin{columns}[options]\begin{column} {widths column 1}...\begin{column}{widths column 2}....\end{columns} and divides a slide vertically into columns. The possible options for the columns command are summarized in Table 4.3.

A simple example of a slide which is structured by a block and a column structure is shown in Listing 4.6 and Fig. 4.4.

```
1  \documentclass{beamer}
2
3  \begin{document}
4
5  \begin{frame}{Block and Column Structures}
6
7      \begin{block}{Hooke's Law}
8          \begin{equation}
9              \sigma=E\varepsilon
10         \end{equation}
11     \end{block}
12
```

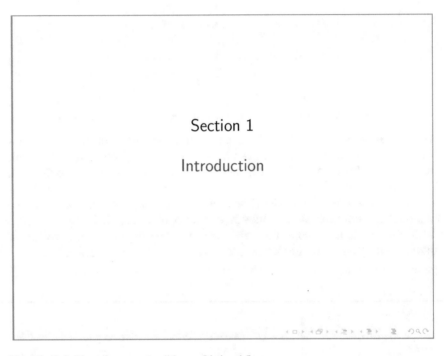

**Fig. 4.3** Definition of a separator slide, see Listing 4.5

**Table 4.3** Different options for the columns environment, see [40]

| Command | Comment |
| --- | --- |
| b | The bottom lines of the columns are vertically aligned |
| c | The columns are centered vertically relative to each other |
| onlytextwidth | The same as totalwidth=\textwidth |
| t | The first lines of the columns are aligned (more specifically, the baselines of the first lines) |
| T | The tops of the first lines of the columns are aligned |
| totalwidth=⟨width⟩ | The columns do not occupy the whole page width |

---

# Block and Column Structures

Hooke's Law

$$\sigma = E\varepsilon \tag{1}$$

Let use have a look on some assumptions and the application of this constitutive relationship.

Assumptions:

► isotropic

► homogeneous

► $E$ constant

Used for common engineering materials in the elastic range, i.e. where a linear relationship in the stress-strain diagram holds.

In the plastic range, we need in addition the following components: a yield condition, a flow rule, and a hardening law.

---

**Fig. 4.4** Structuring a regular slide with a block and column structure, see Listing 4.6

```
13  \noindent Let use have a look on some assumptions and the
14  application of this constitutive relationship.
15
16  \vspace*{5mm}
17
18      \begin{columns}[c]
19          \column{.45\textwidth}
20              Assumptions:
21              \begin{itemize}
22              \item isotropic
23              \item homogeneous
24              \item $E$ constant
25              \end{itemize}
26          \column{.45\textwidth}
27              Used for common engineering materials in the elastic range, i.e.
28              where a linear relationship in the stress–strain diagram holds.
29      \end{columns}
30
31  \vspace*{5mm}
32
33  \noindent In the plastic range, we need in addition the following
34  components: a yield condition, a flow rule, and a hardening law.
35
36  \end{frame}
37
38  \end{document}
```

**Listing 4.6** Structuring a regular slide with a block and column structure, see Fig. 4.4

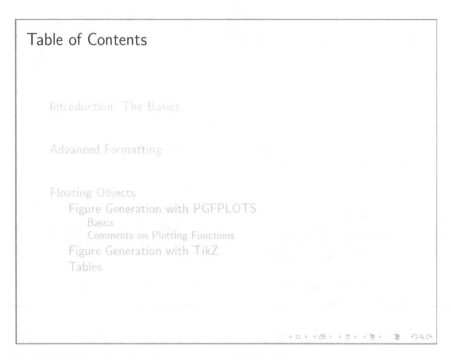

**Fig. 4.5**  Definition of a table of contents page, see Listing 4.7

### 4.1.1.3    Table of Contents, References, Appendix

If a table of contents page should be automatically generated, it is best to structure
the presentation with the \section, \subsection, and \subsubsection commands,
see Sect. 4.1.1.2. Then, the code in Listing 4.7 generates automatically a table of
contents page, see Fig. 4.5.

```
1   \documentclass{beamer}
2
3   \begin{document}
4
5       \begin{frame}{Table of Contents}
6           \tableofcontents[currentsection]
7       \end{frame}
8
9       \section[Intro]{Introduction: The Basics}
10      \begin{frame}{A Simple Frame (Slide)}
11      ...
12
13  \end{document}
```

**Listing 4.7**  Definition of a table of contents page, see Fig. 4.5

The automatic generation of a reference section within a beamer presentation can be
based on the classical \begin{thebibliography}{...}...\end{thebibliography} environment. Then, each bibliographical entry can be initiated by the \bibitem{...} com-

mand. A typical reference section in the Beamer layout is illustrated in Listing 4.8 and Fig. 4.6 whereas a more classical layout, for example as in a book or thesis, is illustrated in Listing 4.9 and Fig. 4.7.

```
1   \documentclass{beamer}
2
3   \begin{document}
4
5   \begin{frame}\frametitle{References}
6
7     \begin{thebibliography}{10}
8
9     \setbeamertemplate{bibliography item}[article]\bibitem{Oe03}
10    \"{O}chsner A, Lamprecht K (2003)
11    \newblock On the uniaxial compression behavior of regular shaped cellular
12             metals.
13    \newblock Mech Res Commun 30:573—-579
14
15    \setbeamertemplate{bibliography item}[book]\bibitem{Oec18}
16    \"{O}chsner A, Merkel M (2018)
17    \newblock One-dimensional finite elements: an introduction to the FE method.
18    \newblock Springer, Berlin
19
20    \setbeamertemplate{bibliography item}[book]\bibitem{Oec20}
21    \"{O}chsner A (2020)
22    \newblock Computational statics and dynamics: an introduction based on
23             the finite element method.
24    \newblock Springer, Singapore
25
```

## References

Öchsner A, Lamprecht K (2003)
On the uniaxial compression behavior of regular shaped cellular
metals.
Mech Res Commun 30:573–579

Öchsner A, Merkel M (2018)
One-dimensional finite elements: an introduction to the FE
method.
Springer, Berlin

Öchsner A (2020)
Computational statics and dynamics: an introduction based on
the finite element method.
Springer, Singapore

**Fig. 4.6** Reference section in the BEAMER layout, see Listing 4.8

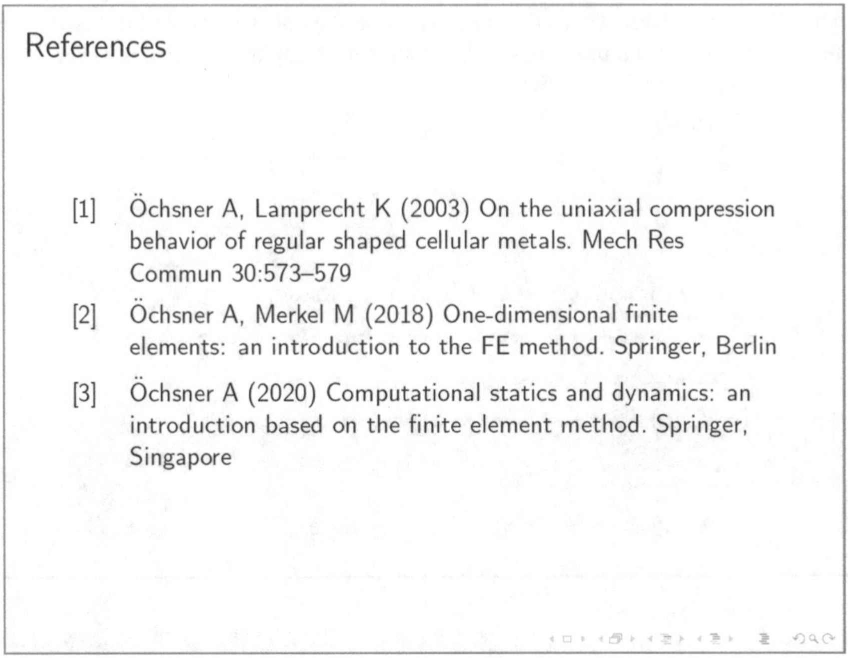

References

[1]   Öchsner A, Lamprecht K (2003) On the uniaxial compression
      behavior of regular shaped cellular metals. Mech Res
      Commun 30:573–579

[2]   Öchsner A, Merkel M (2018) One-dimensional finite
      elements: an introduction to the FE method. Springer, Berlin

[3]   Öchsner A (2020) Computational statics and dynamics: an
      introduction based on the finite element method. Springer,
      Singapore

**Fig. 4.7**  Reference section in the classical layout, see Listing 4.9

```
26      \end{thebibliography}
27   \end{frame}
28   \end{document}
```

**Listing 4.8**  Reference section in the BEAMER layout, see Fig. 4.6

```
1    \documentclass{beamer}
2
3    \begin{document}
4
5    \section{References}
6    \begin{frame}{References}
7
8    \setbeamertemplate{bibliography item}[text]
9
10     \begin{thebibliography}{99.}%
11
12     \bibitem{Oe03}
13     \"{O}chsner A, Lamprecht K (2003) On the uniaxial compression behavior of
14     regular shaped cellular metals. Mech Res Commun 30:573—579
15
16     \bibitem{Oec18}
17     \"{O}chsner A, Merkel M (2018) One-dimensional finite elements: an
18     introduction to the FE method.
19     Springer, Berlin
20
21     \bibitem{Oec20}
22     \"{O}chsner A (2020) Computational statics and dynamics: an introduction
23     based on the finite element method. Springer, Singapore
24
25     \end{thebibliography}
```

```
26
27    \end{frame}
28
29    \end{document}
```

**Listing 4.9** Reference section in the classical layout, see Fig. 4.7

An appendix section is quite common in a book or thesis. In the scope of a presentation, an appendix section might be used to collect additional slides ('backup slides') to answer questions. The declaration of the appendix is done by the \appendix command, see Listing 4.10 for a simple example and Fig. 4.8 for the corresponding output. The use of \appendixname ensures that the section name gets the standard name for an appendix assigned (in this example: 'Appendix'). There is another feature included (see the \setbeamertemplate command) to give the appendix slides their own frame numbering. Without this, the appendix slide would be numbered '3/3' in our example, where it is the third slide of the presentation.

```
1     \documentclass{beamer}
2
3     \usetheme{AnnArbor}
4     \setbeamertemplate{page number in head/foot}[appendixframenumber]
5
6     \begin{document}
7
8         \section{Introduction}
9         \begin{frame}
10            This is the first regular frame
11        \end{frame}
12
13        \begin{frame}
14            This is the second regular frame
15        \end{frame}
16
17        \appendix
18
19        \section{\appendixname}
20        \subsection{Centroids}
21        \begin{frame}
22            This is an appendix frame
23        \end{frame}
24
25    \end{document}
```

**Listing 4.10** Examples of regular and appendix BEAMER frames (slides), see Fig. 4.8

## 4.1.2  Adding Effects to a Presentation (Overlays and Hyperlinks)

A simple way to reveal step-by-step parts of a slide (frame) is the \pause command. Only after pressing the space or return key,[2] the content after the \pause command is displayed. A simple example is shown in Listing 4.11 and Fig. 4.9.

---

[2] Or pressing the page down key, or using the mouse to scroll down or clicking on the next slide button.

**(a)**

**(b)**

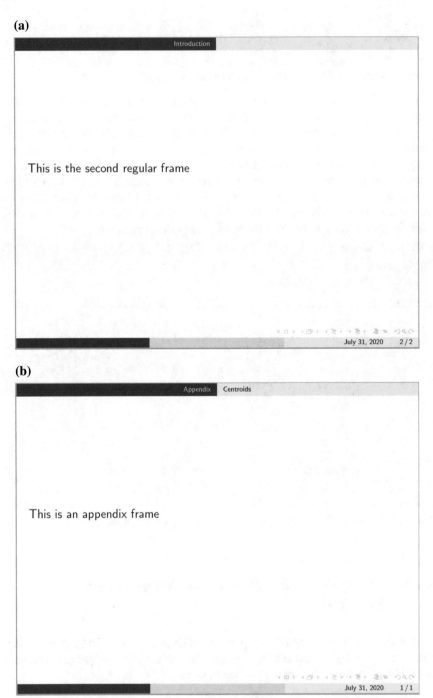

**Fig. 4.8** Examples of BEAMER frames (slides): **a** regular slide, **b** appendix slide (with own frame numbering), see Listing 4.10

**Fig. 4.9** A simple animation based on the \pause command: **a** initial view, **b** including the second part, and **c** including the third part, see Listing 4.11

**(a)**

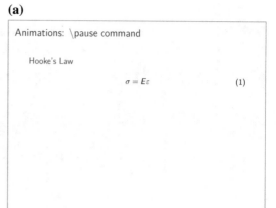

**(b)**

Animations: \pause command

Hooke's Law

$$\sigma = E\varepsilon \qquad (1)$$

Let use have a look on some assumptions and the application of this constitutive relationship.

Assumptions:
- isotropic
- homogeneous
- $E$ constant

Used for common engineering materials in the elastic range, i.e. where a linear relationship in the stress-strain diagram holds.

**(c)**

Animations: \pause command

Hooke's Law

$$\sigma = E\varepsilon \qquad (1)$$

Let use have a look on some assumptions and the application of this constitutive relationship.

Assumptions:
- isotropic
- homogeneous
- $E$ constant

Used for common engineering materials in the elastic range, i.e. where a linear relationship in the stress-strain diagram holds.

In the plastic range, we need in addition the following components: a yield condition, a flow rule, and a hardening law.

```
1   \documentclass{beamer}
2
3   \begin{document}
4
5       \begin{frame}{Animations: \textbackslash pause command}
6
7           \begin{block}{Hooke's Law}
8           \begin{equation}
9           \sigma=E\varepsilon
10          \end{equation}
11          \end{block}
12
13      \pause
14
15      \noindent Let use have a look on some assumptions and the application of this
16          constitutive relationship.
17      \vspace*{5mm}
18
19          \begin{columns}[c]
20          \column{.45\textwidth}
21          Assumptions:
22          \begin{itemize}
23          \item isotropic
24          \item homogeneous
25          \item $E$ constant
26          \end{itemize}
27          \column{.45\textwidth}
28          Used for common engineering materials in the elastic range, i.e.
29              where a linear relationship in the stress-strain diagram holds.
30          \end{columns}
31          \vspace*{5mm}
32
33      \pause
34
35      \noindent In the plastic range, we need in addition the following components:
36      a yield condition, a flow rule, and a hardening law.
37
38          \end{frame}
39  \end{document}
```

**Listing 4.11**  A simple animation based on the \pause command, see Fig. 4.9

A common environment for presentations is the \begin{itemize}...\end{itemize}
environment. The above introduced \pause command can simply be used in this
construct to not reveal the entire list from the beginning. A simple example is shown
in Listing 4.12 and Fig. 4.10 where only the first item of the itemize environment
is shown right from the beginning. The remaining four items are only revealed after
pressing the space or return key

```
1   \documentclass{beamer}
2
3   \begin{document}
4
5   \begin{frame}{Animations: itemize environment}
6
7   \noindent Typical assumptions for Hooke's law in its simplest formulation:
8
9       \begin{itemize}
10          \item  $\sigma=E\varepsilon$
11          \pause
12          \item  $E$ constant
13          \item  elastic range
14          \item  isotropic
15          \item  homogeneous
16      \end{itemize}
```

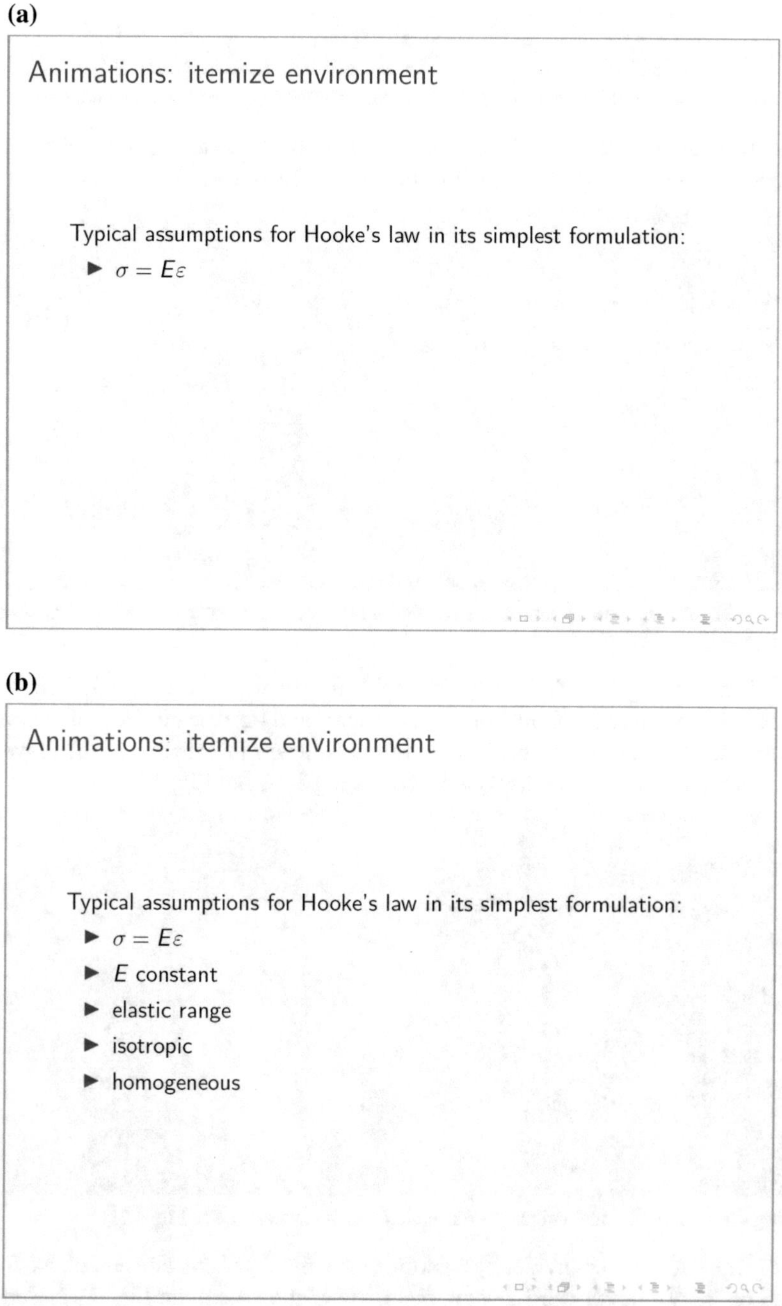

**(a)**

**(b)**

**Fig. 4.10** An animation for the itemize environment based on the \pause command: **a** initial view, **b** including the second part, see Listing 4.12

```
17
18   \end{frame}
19   \end{document}
```

**Listing 4.12**  An animation for the itemize environment based on the \pause command, see Fig. 4.10

If the items of an **itemize** environment should be revealed step-by-step, one can use the option [<+->] of the **itemize** environment, see Listing 4.13.

```
1    \documentclass{beamer}
2
3    \begin{document}
4
5    \begin{frame}{Animations: itemize environment}
6
7    \noindent Typical assumptions for Hooke's law in its simplest formulation:
8
9        \begin{itemize}[<+->]
10           \item   $\sigma=E\varepsilon$
11           \item   $E$ constant
12           \item   elastic range
13           \item   isotropic
14           \item   homogeneous
15       \end{itemize}
16
17   \end{frame}
18   \end{document}
```

**Listing 4.13**  A step-by-step animation for the itemize environment based on the environment option [<+->]

There is also a way to individually customize the display of each single item based on the <i-j> option of the \item command, where i and j are the numbers of the events the particular item is displayed. The option <i-> means that the item is displayed as event number i and it stays displayed, see Listing 4.14 and Fig. 4.11.

```
1    \documentclass{beamer}
2
3    \begin{document}
4
5    \begin{frame}{Animations: itemize environment}
6
7    \noindent Typical assumptions for Hooke's law in its simplest formulation:
8
9        \begin{itemize}
10           \item<1-> $\sigma=E\varepsilon$
11           \item<2-2> $E$ constant
12           \item<3-> elastic range
13           \item<4-> isotropic
14           \item<5-> homogeneous
15       \end{itemize}
16
17   \end{frame}
18   \end{document}
```

**Listing 4.14**  An individual animation for the itemize environment, see Fig. 4.11

There is also a way to individually customize the display of any element on the frame based on the \uncover<i-j>{...} command, see Listing 4.15 and Fig. 4.12. Furthermore, there is in addition the \only<...> command which works as the \uncover command but does not reserve any space on the slide when hidden.

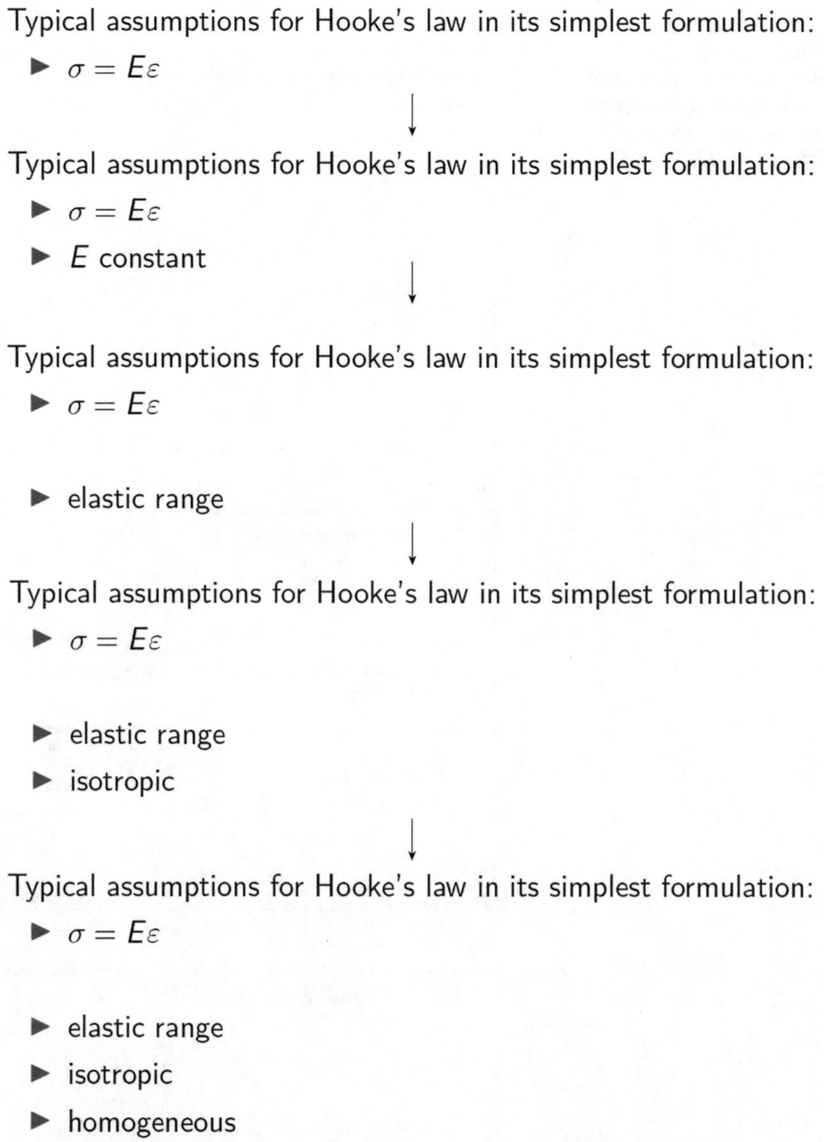

Typical assumptions for Hooke's law in its simplest formulation:

▶ $\sigma = E\varepsilon$

↓

Typical assumptions for Hooke's law in its simplest formulation:

▶ $\sigma = E\varepsilon$

▶ $E$ constant

↓

Typical assumptions for Hooke's law in its simplest formulation:

▶ $\sigma = E\varepsilon$

▶ elastic range

↓

Typical assumptions for Hooke's law in its simplest formulation:

▶ $\sigma = E\varepsilon$

▶ elastic range

▶ isotropic

↓

Typical assumptions for Hooke's law in its simplest formulation:

▶ $\sigma = E\varepsilon$

▶ elastic range

▶ isotropic

▶ homogeneous

**Fig. 4.11**  An individual animation for the itemize environment, see Listing 4.14

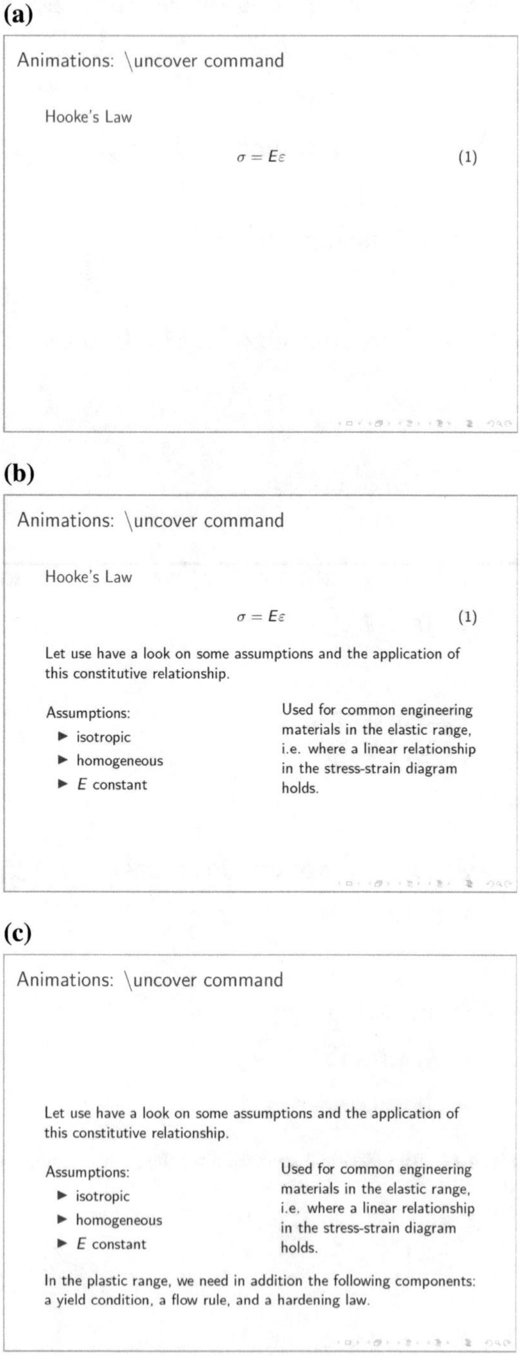

**Fig. 4.12**  An individual
animation based on the
\uncover command: **a** initial
view, **b** including the second
part, and **c** including the
third part, see Listing 4.15

**(a)**

Animations: \uncover command

Hooke's Law

$$\sigma = E\varepsilon \qquad (1)$$

**(b)**

Animations: \uncover command

Hooke's Law

$$\sigma = E\varepsilon \qquad (1)$$

Let use have a look on some assumptions and the application of
this constitutive relationship.

Assumptions:                          Used for common engineering
                                      materials in the elastic range,
  ▶ isotropic                         i.e. where a linear relationship
  ▶ homogeneous                       in the stress-strain diagram
  ▶ $E$ constant                      holds.

**(c)**

Animations: \uncover command

Let use have a look on some assumptions and the application of
this constitutive relationship.

Assumptions:                          Used for common engineering
                                      materials in the elastic range,
  ▶ isotropic                         i.e. where a linear relationship
  ▶ homogeneous                       in the stress-strain diagram
  ▶ $E$ constant                      holds.

In the plastic range, we need in addition the following components:
a yield condition, a flow rule, and a hardening law.

```
1   \documentclass{beamer}
2
3   \begin{document}
4
5   \begin{frame}{Animations: uncover command}
6
7   \uncover<1-2>
8       {
9       \begin{block}{Hooke's Law}
10      \begin{equation}
11      \sigma=E\varepsilon
12      \end{equation}
13      \end{block}
14      }
15
16  \uncover<2->
17      {
18      \noindent Let use have a look on some assumptions and the application of
19       this constitutive  relationship.
20      \vspace*{5mm}
21
22      \begin{columns}[c]
23      \column{.45\textwidth}
24      Assumptions:
25      \begin{itemize}
26      \item isotropic
27      \item homogeneous
28      \item $E$ constant
29      \end{itemize}
30      \column{.45\textwidth}
31      Used for common engineering materials in the elastic range, i.e. where a
32       linear relationship in the stress-strain diagram holds.
33      \end{columns}
34      }
35
36      \vspace*{5mm}
37
38  \uncover<3->
39      {
40      \noindent In the plastic range, we need in addition the following
41       components: a yield condition, a flow rule, and a hardening law.
42      }
43
44  \end{frame}
45
46  \end{document}
```

**Listing 4.15**  An individual animation based on the \uncover command, see Fig. 4.12

In case that several figures should be overlaid in the same float environment, attention must be paid to changing figure labels. A practical way is the use of the \only command as illustrated in Listing 4.16 and Fig. 4.13.

```
1   \documentclass{beamer}
2
3   \begin{document}
4
5   \begin{frame}{Animations: overlay of figures}
6
7   \begin{figure}[h!]
8       \includegraphics<1>[scale=0.04]{figs/tensile_direction.jpg}
9       \includegraphics<2>[scale=0.04]{figs/compressive_direction.jpg}
10      \caption{\only<1>{Build orientation of tensile sampless}
11               \only<2>{Build orientation of compression samples}}
12      label{fig:printed_samples}
13  \end{figure}
```

**(a)**

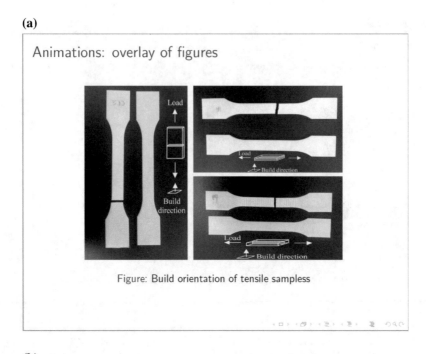

**(b)**

**Fig. 4.13** Overlay of figures in the same float environment: **a** first set of figure and caption, **b** second set of figure and caption, see Listing 4.16

```
14
15    \end{frame}
16
17  \end{document}
```

**Listing 4.16**  Overlay of figures in the same float environment, see Fig. 4.13

The incorporation of hyperlinksrequires the initialization of the LaTeX package hyperref in the preamble of the document, see Listing 4.17. The location to jump to
can be defined by the \hypertarget<overlay specification>{ref. name}{text} command or it can simply be a \label. The hyperlink can be formed by a particular button based on a combination of the \hyperlink{ref. name}{link text} and
\beamergotobutton{button text} commands, see Listing 4.17 and Fig. 4.14 for
the graphical output.

```
1   \documentclass{beamer}
2   \usepackage{hyperref}
3
4   \begin{document}
5
6       \begin{frame}[label=first_slide]
7           \begin{itemize}
8               \item<1-> First item.
9               \item<2-> Second item.
10              \item<3-> Third item.
11          \end{itemize}
12
13          \hyperlink{third_item}{\beamergotobutton{Jump to third slide}}
14          \hypertarget<3>{third_item}{}
15          \hyperlink{fig:beam_gen}{\beamergotobutton{Jump to
16                              Fig.~\ref{fig:beam_gen}}}
17      \end{frame}
18
19      \begin{frame}
20          \begin{figure}[h!]
21          \centering
22          \includegraphics[scale=0.045]{Figs/beam_gen.jpg}
23          \caption{Schematic representation of a thin beam}
24          \label{fig:beam_gen}        % Give a unique label
25          \end{figure}
26
27          \hyperlink{first_slide}{\beamergotobutton{Jump back to slide~1}}\\
28      \end{frame}
29
30  \end{document}
```

**Listing 4.17**  Including hyperlinks, see Fig. 4.14

The \hypertarget and \hyperlink construct in Listing 4.17 can be replaced by introducing a label for the first slide (in our example called 'first_slide', see line 6) and
referring to it within the \hyperlink command (line 13), see Listing 4.18. Furthermore, this example uses another standard button, the so-called return button, see
line 26.

```
1   \documentclass{beamer}
2   \usepackage{hyperref}
3
4   \begin{document}
5
6       \begin{frame}[label=first_slide]
7           \begin{itemize}
8               \item<1-> First item.
```

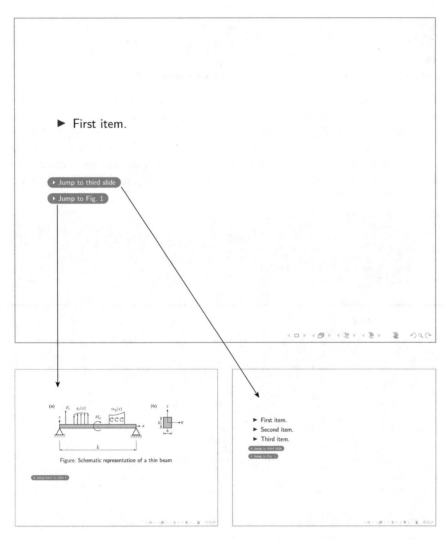

**Fig. 4.14** Including hyperlinks, see Listing 4.17

```
 9              \item<2-> Second item.
10              \item<3-> Third item.
11          \end{itemize}
12
13      \hyperlink{first_slide<3>}{\beamergotobutton{Jump to third slide}}
14      \hyperlink{fig:beam_gen}{\beamergotobutton{Jump to
15                               Fig.~\ref{fig:beam_gen}}}
16      \end{frame}
17
18      \begin{frame}
19          \begin{figure}[h!]
20          \centering
21          \includegraphics[scale=0.045]{Figs/beam_gen.jpg}
```

```
22        \caption{Schematic representation of a thin beam}
23        \label{fig:beam_gen}          % Give a unique label
24        \end{figure}
25
26      \hyperlink{first_slide}{\beamerreturnbutton{Return}}
27      \end{frame}
28
29    \end{document}
```

**Listing 4.18**  Including hyperlinks, modifications

The beamer class offers many more pre-defined buttons and hyperlink commands for navigation purposes and Table 4.4 summarizes some of the available options.

In order to change the layout of the displayed buttons, one can use the commands in Table 4.5.

**Table 4.4**  Pre-defined buttons and hyperlink commands for navigation purpose, see [40]

| Command | Comment |
| --- | --- |
| \beamergotobutton{button text} | To jump to a hyperlink or to a label |
| \beamerskipbutton{button text} | To skip over a certain part |
| \beamerreturnbutton{button text} | To return to a hyperlink or to a label |
| \hyperlink<overlay specification>{ref. name}{link text} | To jump to the slide on which a hypertarget or label was used as 'ref. name' |
| \hyperlinkslideprev<overlay specification>{link text} | To jump one slide back |
| \hyperlinkslidenext<overlay specification>{link text} | To jump one slide forward |
| \hyperlinkframestart<overlay specification>{link text} | To jump to the first slide of the current frame |
| \hyperlinkframeend<overlay specification>{link text} | To jump to the last slide of the current frame |
| \hyperlinkframestartnext<overlay specification>{link text} | To jump to the first slide of the next frame |
| \hyperlinkframeendprev<overlay specification>{link text} | To jump to the last slide of the previous frame |
| \hyperlinkpresentationstart<overlay specification>{link text} | To jump to the first slide of the presentation |
| \hyperlinkpresentationend<overlay specification>{link text} | To jump to the last slide of the presentation (excluding any appendix) |
| \hyperlinkappendixstart<overlay specification>{link text} | To jump to the first slide of the appendix |
| \hyperlinkappendixend<overlay specification>{link text} | To jump to the last slide of the appendix |
| \hyperlinkdocumentstart<overlay specification>{link text} | To jump to the first slide of the presentation |
| \hyperlinkdocumentend<overlay specification>{link text} | To jump to the last slide of the presentation (including the appendix) |

### 4.1.3  Including Graphics and Tables

To include an external graphic file (e.g. in EPS, PDF, PNG, or JPG format), one can use the classical LATEX command \includegraphics[options]{filename.eps}, see [9] for details. As in any other LATEX document format, this command can be embedded in the \begin{figure}[option]...\end{figure} environment to create a floating object with the possibility of a figure caption and label. A simple example to include an external JPG figure (the same procedure holds for a PNG or PDF file) is shown in Listing 4.19.

```
1   \documentclass{beamer}
2
3   \begin{document}
4
5       \begin{frame}{Including a JPG figure}
6           \begin{figure}[h!]
7               \centering
8               \includegraphics[scale=0.045]{Figs/beam_gen.jpg}
9               \caption{Schematic representation of a thin beam}
10              \label{fig:beam_gen}
11          \end{figure}
12      \end{frame}
13
14  \end{document}
```

**Listing 4.19**  Including a JPG figure

The incorporation of an external EPS file may require to include the LATEX package graphicx and epstopdf (i.e., to convert the EPS file into the PDF format), see Listing 4.20 and Fig. 4.15 for the graphical output.

```
1   \documentclass{beamer}
2
3   \usepackage{graphicx}
4   \usepackage[outdir=./PDF/]{epstopdf}
5
6   \begin{document}
7
8       \begin{frame}{Including an EPS figure}
9           \begin{figure}[h!]
10              \centering
11              \includegraphics[scale=0.70]{Figs/beam_gen.eps}
12              \caption{Schematic representation of a thin beam}
13              \label{fig:beam_gen}
14          \end{figure}
15      \end{frame}
16
17  \end{document}
```

**Listing 4.20**  Including an EPS figure, see Fig. 4.15

The default setting does not produce any figure number as shown in Fig. 4.15. If this is required, the command \setbeamertemplate{caption}[numbered] changes this default setting. Furthermore, the wording 'Figure' as well as the caption font size can be easily adjusted, see Listing 4.21 and Fig. 4.16 for a simple example.

```
1   \documentclass{beamer}
2
3   \usepackage{graphicx}
4   \usepackage[outdir=./PDF/]{epstopdf}
5
```

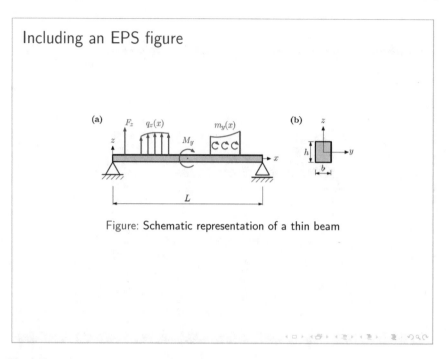

Including an EPS figure

Figure: Schematic representation of a thin beam

**Fig. 4.15** Including an EPS figure, see Listing 4.20

```
6    \setbeamertemplate{caption}[numbered]
7    \setbeamerfont{caption}{size=\scriptsize}
8    \renewcommand{\figurename}{Fig.}.
9
10   \begin{document}
11
12       \begin{frame}{Including an EPS figure, modified caption style}
13           \begin{figure}[h!]
14               \centering
15               \includegraphics[scale=0.70]{Figs/beam_gen.eps}
16               \caption{Schematic representation of a thin beam}
17               \label{fig:beam_gen}
18           \end{figure}
19       \end{frame}
20
21   \end{document}
```

**Listing 4.21** Including an EPS figure and modified caption style, see Fig. 4.16

Figures directly generated based on the TikZ package as outlined in Sect. 3.2 can be directly incorporated as a floating object. The corresponding TikZ package must be included in the preamble of a document. A simple example is shown (see the original Listing 3.70) in Listing 4.22, see Fig. 4.17 for the graphical output.

```
1    \documentclass{beamer}
2
3    \usepackage{tikz}
4
5    \begin{document}
```

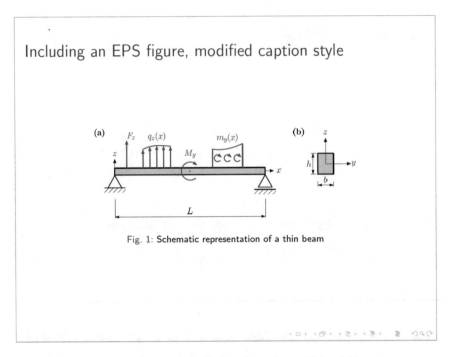

Fig. 1: Schematic representation of a thin beam

**Fig. 4.16** Including an EPS figure and modified caption style, see Listing 4.21

```
6
7    \begin{frame}{Including a TikZ picture}
8        \begin{figure}[h!]
9        \centering
10       \begin{tikzpicture}
11           \begin{scope}[rotate=30]
12           \draw (3,3) — (5,4);
13           \draw (3,3) circle (0.75);
14           \end{scope}
15       \end{tikzpicture}
16       \caption{Grouping of graphical objects and common
17                   transformation (rotation)}
18       \label{fig:scope_o}
19       \end{figure}
20   \end{frame}
21
22 \end{document}
```

**Listing 4.22** Including a TikZ picture, see Fig. 4.17

A table can be handled as any other floating object and Listing 4.23 shows a simple implementation (see Fig. 4.18 for the graphical output).

```
1  \documentclass{beamer}
2
3  \begin{document}
4
5  \begin{frame}{Including a table}
6  \begin{table}
7  \centering
```

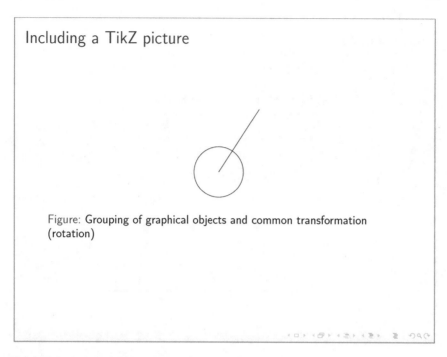

**Fig. 4.17** Including a TikZ picture, see Listing 4.22

```
8    \caption{Integration rules for plane elasticity elements}
9    \label{tab:prelude_int_plane}        % Give a unique label
10       \begin{tabular}{cccccc}
11       \hline\noalign{\smallskip}
12       Points& & \multicolumn{1}{c}{$\xi_i$} & \multicolumn{1}{c}{$\eta_i$}
13          & \multicolumn{1}{c}{Weight $w_i$} \\
14       \noalign{\smallskip}\hline\noalign{\smallskip}
15       1  & & $0$        & $0$        & $4$          \\[1.5ex]
16       4  & & $\pm 1/\sqrt{3}$ & $\pm 1/\sqrt{3}$ & $1$       \\[1.5ex]
17       9  & & $0$        & $0$        & $\tfrac{64}{81}$   \\[1.5ex]
18          & & $0$        & $\pm \sqrt{6}$  & $\tfrac{40}{81}$   \\[1.5ex]
19          & & $\pm \sqrt{6}$  & $0$        & $\tfrac{40}{81}$   \\[1.5ex]
20          & & $\pm \sqrt{6}$  & $\pm \sqrt{6}$  & $\tfrac{25}{81}$   \\[0.5ex]
21       \noalign{\smallskip}\hline\noalign{\smallskip}
22       \end{tabular}
23    \end{table}
24    \end{frame}
25
26    \end{document}
```

**Listing 4.23** Including a table, see Fig. 4.18

Some users may have already a LATEXdocument such as a manuscript or a thesis and must prepare then a presentation. Transferring floating objects from one document class to the beamer class may not result at first in the desired outcome since the size of the floating object requires some adjustment in the beamer document. The following Listings 4.24–4.26 explain the scaling of floating TikZ pictures, included pictures and tables.

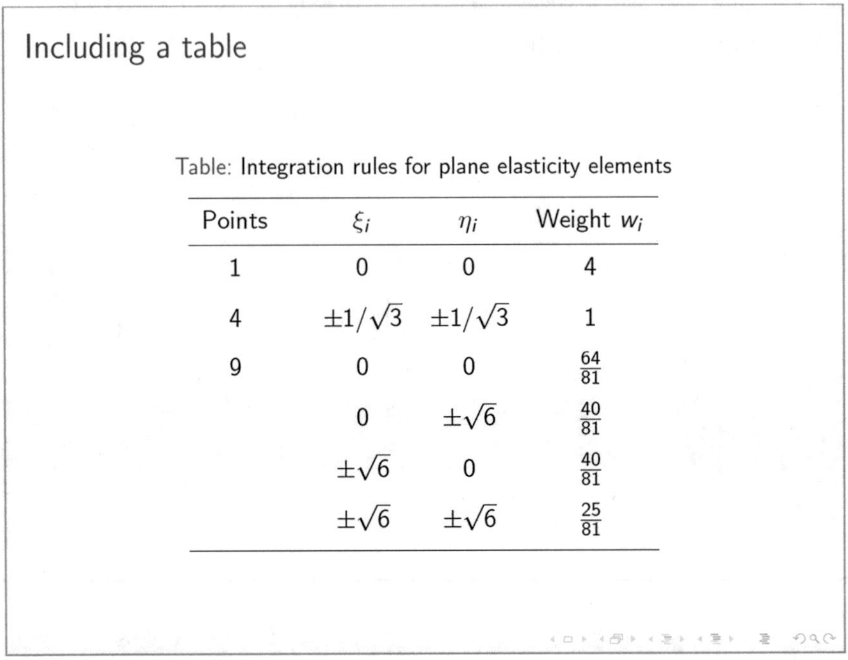

## Including a table

Table: Integration rules for plane elasticity elements

| Points | $\xi_i$ | $\eta_i$ | Weight $w_i$ |
|--------|---------|----------|--------------|
| 1 | 0 | 0 | 4 |
| 4 | $\pm 1/\sqrt{3}$ | $\pm 1/\sqrt{3}$ | 1 |
| 9 | 0 | 0 | $\frac{64}{81}$ |
| | 0 | $\pm\sqrt{6}$ | $\frac{40}{81}$ |
| | $\pm\sqrt{6}$ | 0 | $\frac{40}{81}$ |
| | $\pm\sqrt{6}$ | $\pm\sqrt{6}$ | $\frac{25}{81}$ |

**Fig. 4.18**  Including a table, see Listing 4.23

```
1  \begin{figure}
2    \centering
3    \begin{tikzpicture}[scale=0.50,transform shape]
4    ...
5    \end{tikzpicture}
6    \caption{...}
7    \label{...}
8  \end{figure}
```
**Listing 4.24**  Scaling of a floating TikZ picture

```
1  \begin{figure}[h!]
2    \centering
3    \includegraphics[scale=0.50]{Figs/rod_gen.eps}
4    \caption{...}
5    \label{...}
6  \end{figure}
```
**Listing 4.25**  Scaling of an included figure

```
1  \begin{table}
2    \scriptsize
3    \caption{...}
4    \label{...}
5    \begin{tabular}{ll}
6    ...
7    \end{tabular}
8  \end{table}
```
**Listing 4.26**  Scaling of a floating table

The scaling option \scriptsize in Listing 4.26 can be also replaced by the other built-in font sizes such as \tiny or \footnotesize.

### 4.1.4  Including Sounds and Videos

The recommended way of including sounds and videos into a beamer presentation is based on the LaTeXpackage media9 [17]. This package allows to include interactive Flash (SWF) objects, 3D objects (Adobe U3D & PRC), as well as sound and video files (MP3, FLV and MP4 formats). However, a media player Flash component, i.e., commonly the Adobe Flash Player, is required.[3] The older movie15 package should no longer be used [16]. The media9 package provides a simple player for audio files (APlayer.swf), a simple player for video files (VPlayer.swf), and a highly configurable open-source player (StrobeMediaPlayback.swf).

Listing 4.27 shows a simple example of the integration of a MP3 sound file into a beamer presentation [17], see Fig. 4.19 for the graphical output. The basic command to include an audio or video is \includemedia[options]{text}{main Flash (SWF) file}. The generated button (based on the \framebox command) allows to start the audio file. Any further operation, e.g. pause or rewind, must be done via the right-click menu.

```
1   \documentclass{beamer}
2   \usepackage{media9}
3
4   \begin{document}
5
6   \begin{frame}
7
8     \includemedia[
9       addresource=music.mp3,
10      flashvars={
11      source=music.mp3},
12      transparent,
13      passcontext %show player's right-click menu
14      ]{\color{blue}\framebox[0.3\linewidth][c]{MP3 music}}{APlayer.swf}
15
16   \end{frame}
17
18   \end{document}
```

**Listing 4.27**  Simple example of an embedded MP3 sound file, see Fig. 4.19

A more advanced example of the integration of a MP3 sound file into a beamer presentation is shown in Listing 4.28, see Fig. 4.20 for the graphical representation. This example features the creation of further buttons (based on the commands \mediabutton and \beamerbutton) to easier operate the MP3 file.

```
1   \documentclass{beamer}
2   \usepackage{media9}
3
4   \begin{document}
```

---

[3] Modifications will be required in the future because the use of the Flash Player will no longer be supported by Adobe Acrobat after December 2020.

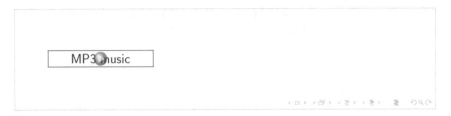

**Fig. 4.19** Simple example of an embedded MP3 sound file, see Listing 4.27

**Fig. 4.20** Advanced example of an embedded MP3 sound file: application of beamer buttons, see Listing 4.28

**Table 4.5** Some commands to change the appearance of BEAMER buttons

| Command | Comment |
|---|---|
| \setbeamercolor{button}{bg=orange,fg=violet} | Color of the background (bg) and the foreground (fg), i.e. font |
| \setbeamerfont{button}{size=\tiny,shape= \itshape,family=\rmfamily} | Size, shape and family of the used font |

```
5
6    \begin{frame}
7
8      \begin{center}
9        \begin{scriptsize}\textcolor{blue}{MP3 Audio:}\;
10       \includemedia[label=show4,addresource=Audio/music.mp3,
11       flashvars={source=Audio/music.mp3},
12       ]{\beamerbutton{play}}{APlayer.swf}
13       \end{scriptsize}
14       \mediabutton[mediacommand=show4:pause]{\beamerbutton{pause}}
15       \mediabutton[mediacommand=show4:rewind]{\beamerbutton{rewind}}
16     \end{center}
17
18   \end{frame}
19
20 \end{document}
```

**Listing 4.28** Advanced example of an embedded MP3 sound file: application of beamer buttons, see Fig. 4.20

In order to change the layout of the displayed buttons in Fig. 4.20, one can use the provided commands in Table 4.5.

The final example shows the integration of a MP4 video file into a beamer presentation, see Listing 4.29 and Fig. 4.21. Attention must be paid to the size of the framebox, which should be defined in the size of the video in the beamer presentation. This task

**Fig. 4.21**  Simple example of an embedded MP4 video file, see Listing 4.29

can be achieved by the \framebox(rectangular width in pt,rectangular height in pt){text}
command.

```
1  \documentclass{beamer}
2  \usepackage{media9}
3
4  \begin{document}
5
6   \begin{frame}
7
8      \begin{center}
9       \includemedia[
10        addresource=video.mp4,flashvars={source=video.mp4
11        &autoPlay=true},
12        transparent,passcontext]
13        {\color{blue}\framebox(150,90){MP4 video}}{VPlayer.swf}
14     \end{center}
15
16  \end{frame}
17
18  \end{document}
```

**Listing 4.29**  Simple example of an embedded MP4 video file, see Fig. 4.21

### 4.1.5  Customizing the Layout by Built-In Themes

The layout of a BEAMER presentation can easily be modified by relying on built-in
themes, i.e. predefined settings of the layout. Such a theme can be initiated by the
\usetheme{...} command in the preamble of the document, see Listing 4.30. There is
a variety of official BEAMER themes available, see Table 4.6. An informative overview
on the layout of the different themes is provided by the web pages [7, 8].

```
1  \documentclass{beamer}
2  \usetheme{default}
3
4  \title{Example Presentation Created with the Beamer Class}
5  \author{John Smith}
```

**Table 4.6** Official BEAMER presentation themes. For a graphical overview, see [7, 8]

| AnnArbor | Antibes | Bergen | Berkeley | Berlin | Boadilla |
|----------|---------|--------|----------|--------|----------|
| boxes | CambridgeUS | Copenhagen | Darmstadt | default | Dresden |
| Frankfurt | Goettingen | Hannover | Ilmenau | JuanLesPins | Luebeck |
| Madrid | Malmoe | Marburg | Montpellier | PaloAlto | Pittsburgh |
| Rochester | Singapore | Szeged | Warsaw | | |

```
6    \date{\today}
7
8    \begin{document}
9
10       \begin{frame}
11           \titlepage
12       \end{frame}
13
14   \end{document}
```

**Listing 4.30** Customizing the layout by built-in themes

A comparative example of three different BEAMER themes, i.e., default, Berlin and Berkeley, is shown in Fig. 4.22.

The predefined BEAMER presentation themes can be further customized based on color themes, outer themes, inner themes, and font themes. Let us look first on the so-called color themes, which must be initiated by the \usecolortheme{...} command in the preamble of the document, see Listing 4.31. There is a variety of official BEAMER color themes available, see Table 4.7.

```
1    \documentclass{beamer}
2    \usetheme{Berlin}
3    \usecolortheme{wolverine}
4
5    \title{Example Presentation Created with the Beamer Class}
6    \author{John Smith}
7    \date{\today}
8
9    \begin{document}
10
11       \begin{frame}
12           \titlepage
13       \end{frame}
14
15   \end{document}
```

**Listing 4.31** Initialization of a BEAMER color theme in the preamble

A comparative example of three different BEAMER color themes, i.e., default, beaver and wolverine, is shown in Fig. 4.23.

The layout of the border of a frame, i.e., definitions with regards to head- and footlines, sidebar, logo, navigation symbols and bars, can be customized based on so-called outer themes, which must be initiated by the \useoutertheme{...} command in the preamble of the document, see Listing 4.32. There is a variety of official BEAMER outer themes available, see Table 4.8.

**Fig. 4.22** Examples of different BEAMER presentation themes: **a** default, **b** Berlin, and **c** Berkeley

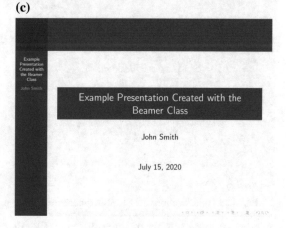

**Table 4.7**  Official BEAMER color themes. For a graphical overview, see [7, 8]

| albatross | beaver | beetle | crane | default | dolphin |
|-----------|--------|--------|-------|---------|---------|
| dove | fly | lily | orchid | rose | seagull |
| seahorse | whale | wolverine | | | |

```
1   \documentclass{beamer}
2   \usetheme{Berlin}
3   \usecolortheme{wolverine}
4
5   \title{Example Presentation Created with the Beamer Class}
6   \author{John Smith}
7   \date{\today}
8
9   \begin{document}
10
11      \begin{frame}
12          \titlepage
13      \end{frame}
14
15  \end{document}
```

**Listing 4.32**  Initialization of a BEAMER color theme in the preamble

A comparative example of three different BEAMER outer themes, i.e., default, info-lines and sidebar, is shown in Fig. 4.24.

The layout of the inner part of a frame, i.e., definitions with regards to enumerations, itemize environments, block environments, theorem environments, or the table of contents, can be customized based on so-called inner themes, which must be initiated by the **\useinnertheme{...}** command in the preamble of the document, see Listing 4.33. The different official BEAMER inner themes are summarized see Table 4.9.

```
1   \documentclass{beamer}
2   \usetheme{Berlin}
3   \useinnertheme{rectangles}
4
5   \begin{document}
6
7       \begin{frame}
8           \begin{enumerate}
9               \item homogeneous materials,
10              \item iostropic materials
11          \end{enumerate}
12
13          \begin{block}{Hooke's Law}
14              \begin{equation}
15                  \sigma = E \times \varepsilon
16              \end{equation}
17          \end{block}
18      \end{frame}
19
20  \end{document}
```

**Listing 4.33**  Initialization of a BEAMER inner theme in the preamble

A comparative example of three different BEAMER outer themes, i.e., rectangles, circles and inmargin, is shown in Fig. 4.25.

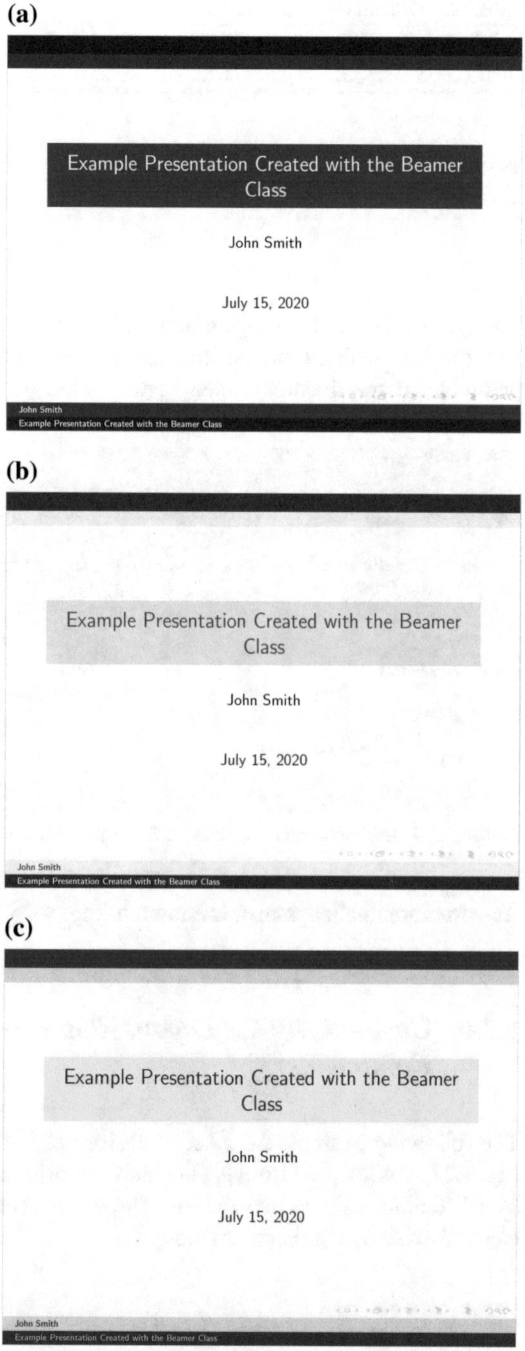

**Fig. 4.23** Examples of different BEAMER color themes: **a** default, **b** beaver, and **c** wolverine

**Table 4.8**  Official BEAMER outer themes

| infolines  | default | miniframes | shadow | sidebar | smoothbars |
|------------|---------|------------|--------|---------|------------|
| smoothtree | split   | tree       |        |         |            |

**Table 4.9**  Official BEAMER inner themes

| circles | default | inmargin | rectangles | rounded |
|---------|---------|----------|------------|---------|

The type of font and corresponding attributes can be customized based on so-called font themes, which must be initiated by the \usefonttheme{...} command in the preamble of the document, see Listing 4.34. The different official BEAMER font themes are summarized see Table 4.10.

```
1   \documentclass{beamer}
2
3   \usetheme{Berlin}
4   \usefonttheme{serif}
5
6   \title{Example Presentation Created with the Beamer Class}
7   \author{John Smith}
8   \date{\today}
9
10
11  \begin{document}
12
13      \begin{frame}
14          \titlepage
15      \end{frame}
16
17  \end{document}
```

**Listing 4.34**  Initialization of a BEAMER font theme in the preamble

A comparative example of three different BEAMER font themes, i.e., default, serif and structuresmallcapsserif, is shown in Fig. 4.26.

### 4.1.6  Customizing the Layout Element-wise by Single Commands

The following example for a set of customized lecture slides, see Listing 4.35 and Fig. 4.27, was adopted from [41], which was originally provided as a lecture template by Till Tantau. It shows how the entire layout of a beamer presentation can be modified element-wise by single commands.

**Fig. 4.24** Examples of different BEAMER outer themes (all three versions use the presentation theme 'Berlin'): **a** default, **b** infolines, and **c** sidebar

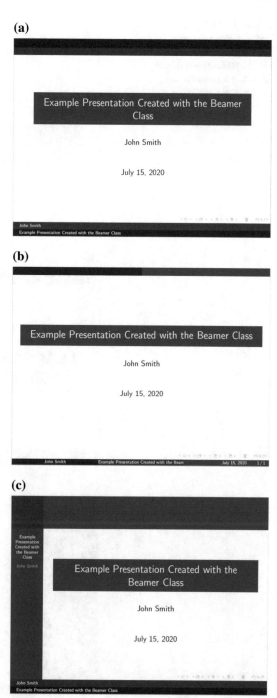

Fig. 4.25 Examples of
different BEAMER inner
themes (all three versions
use the presentation theme
'Berlin'): **a** rectangles, **b**
circles, and **c** inmargin

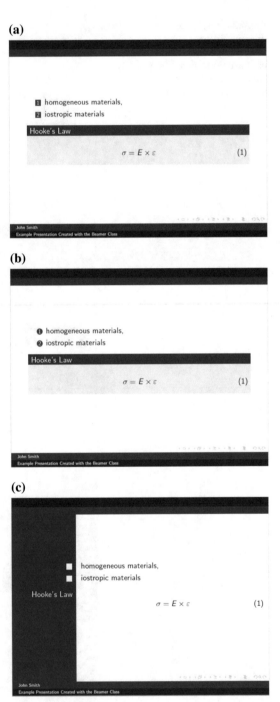

```
1    \documentclass{beamer}
2
3     \usepackage{times}
4     \mode<article>
5     {
6          \usepackage{times}
7          \usepackage{mathptmx}
8          \usepackage[left=1.5cm, right=6cm, top=1.5cm, bottom=3cm]{geometry}
9     }
10
11    \usepackage{hyperref}
12    \usepackage[T1]{fontenc}
13    \usepackage{tikz}
14    \usepackage{mathtools}
15
16    %%% Beamer Version Theme Settings
17
18    \useoutertheme[height=0pt, width=2cm, right]{sidebar}
19    \usecolortheme{rose, sidebartab}
20    \useinnertheme{circles}
21    \usefonttheme[only large]{structurebold}
22
23    \setbeamercolor{sidebar right}{bg=black!15}
24    \setbeamercolor{structure}{fg=blue}
25    \setbeamercolor{author}{parent=structure}
26
27    \setbeamerfont{title}{series=\normalfont, size=\LARGE}
28    \setbeamerfont{title in sidebar}{series=\bfseries}
29    \setbeamerfont{author in sidebar}{series=\bfseries}
30    \setbeamerfont*{item}{series=}
31    \setbeamerfont{frametitle}{size=}
32    \setbeamerfont{block title}{size=\small}
33    \setbeamerfont{subtitle}{size=\normalsize, series=\normalfont}
34
35    \setbeamertemplate{navigation symbols}{}
36    \setbeamertemplate{bibliography item}[book]
37    \setbeamertemplate{sidebar right}
38    {
39         {\usebeamerfont{title in sidebar}%
40             \vskip1.5em%
41             \hskip3pt%
42             \usebeamercolor[fg]{title in sidebar}%
43             \insertshorttitle[width=2cm-6pt, center, respectlinebreaks]\par%
44             \vskip1.25em%
45        }%
46        {%
47             \hskip3pt%
48             \usebeamercolor[fg]{author in sidebar}%
49             \usebeamerfont{author in sidebar}%
50             \insertshortauthor[width=2cm-2pt, center, respectlinebreaks]\par%
51             \vskip1.25em%
52        }%
53        \hbox to2cm{\hss\insertlogo\hss}
54        \vskip1.25em%
55        \insertverticalnavigation{2cm}%
56        \vfill
57        \hbox to 2cm{\hfill\usebeamerfont{subsection in
58            sidebar}\strut\usebeamercolor[fg]{subsection in
59            sidebar}\insertshortlecture.\insertframenumber\hskip5pt}%
60        \vskip3pt%
61    }%
62
63    \setbeamertemplate{title page}
64    {
65        \vbox{}
66        \vskip1em
67        {\huge Lecture \insertshortlecture\par}
```

```
68      {\usebeamercolor[fg]{title}\usebeamerfont{title}\inserttitle\par}%
69      \ifx\insertsubtitle\@empty%
70      \else%
71      \vskip0.25em%
72      {\usebeamerfont{subtitle}\usebeamercolor[fg]{subtitle}
73        \insertsubtitle\par}%
74      \fi%
75      \vskip1em\par
76      Lecture \emph{\lecturename}\ from \insertdate\par
77      \vskip0pt plus1filll
78      \leftskip=0pt plus1fill\insertauthor\par
79      \insertinstitute\vskip1em
80  }
81
82  \logo{\includegraphics[width=1.9cm]{Figs/Esslingen_University_Logo.png}}
83
84  %%% Article Version Layout Settings
85
86  \mode<article>
87
88  \makeatletter
89  \def\@listI{\leftmargin\leftmargini
90      \parsep 0pt
91      \topsep 5\p@    \@plus3\p@  \@minus5\p@
92      \itemsep0pt}
93  \let\@listi=\@listI
94
95
96  \setbeamertemplate{frametitle}{\paragraph*{\insertframetitle\
97      \ \small\insertframesubtitle}\ \par
98  }
99  \setbeamertemplate{frame end}%
100      \marginpar{\scriptsize\hbox to 1cm{\sffamily%
101          \hfill\strut\insertshortlecture.\insertframenumber}\hrule height .2pt}}
102  \setlength{\marginparwidth}{1cm}
103  \setlength{\marginparsep}{4.5cm}
104
105  \def\@maketitle{\makechapter}
106
107  \def\makechapter{
108      \newpage
109      \null
110      \vskip 2em%
111      {%
112          \parindent=0pt
113          \raggedright
114          \sffamily
115          \vskip8pt
116          {\fontsize{36pt}{36pt}\selectfont Lecture \insertshortlecture
117           \par\vskip2pt}
118          {\fontsize{24pt}{28pt}\selectfont \color{blue!50!black}\\
119            \insertlecture\par\vskip4pt}
120          {\Large\selectfont \color{blue!50!black} \insertsubtitle\par}
121          \vskip10pt
122      }
123      \par
124      \vskip 1.5em%
125  }
126
127  \let\origstartsection=\@startsection
128  \def\@startsection#1#2#3#4#5#6{%
129      \origstartsection{#1}{#2}{#3}{#4}{#5}{#6\normalfont\sffamily
130        \color{blue!50!black}\selectfont}}
131
132  \makeatother
133
134  \mode
```

```
135     <all>
136
137     %%% Typesetting Listings
138
139     \usepackage{listings}
140     \lstset{language=Java}
141
142     \alt<presentation>
143     {\lstset{%
144             basicstyle=\footnotesize\ttfamily,
145             commentstyle=\slshape\color{green!50!black},
146             keywordstyle=\bfseries\color{blue!50!black},
147             identifierstyle=\color{blue},
148             stringstyle=\color{orange},
149             escapechar=\#,
150             emphstyle=\color{red}}
151     }
152     {
153         \lstset{%
154             basicstyle=\ttfamily,
155             keywordstyle=\bfseries,
156             commentstyle=\itshape,
157             escapechar=\#,
158             emphstyle=\bfseries\color{red}
159     }
160     }
161
162
163     %%% Define General Slide Details
164
165     \def\lecturename{Finite Element Method}
166     \title{\insertlecture}
167     \author{Prof. Dr.-Ing. Andreas \"{O}chsner}
168     \institute{\begin{footnotesize}
169         University of Applied Sciences Esslingen,\\
170         Faculty of Mechanical Engineering, \\
171         Kanalstr. 33, 73728 Esslingen
172     \end{footnotesize}}
173     \lecture[5]{Thin Beam Element}
174
175     \subtitle{FEM}
176     \date{Monday, 6th July 2020}
177
178     %%% Main Document
179
180     \begin{document}
181
182         \begin{frame}
183             \maketitle
184         \end{frame}
185
186     \section{General}
187
188         \begin{frame}{First Slide Content}
189             \begin{figure}[h!]
190                 \centering
191                 \includegraphics[scale=0.040]{Figs/beam_gen.jpg}
192                 \caption{Schematic representation of a thin beam}
193                 \label{fig:beam_gen}        % Give a unique label
194             \end{figure}
195
196             \begin{block}{this is a block structure}
197                 \begin{enumerate}
198                 \item Text ...
199                 \item Comment ...
200                 \end{enumerate}
201             \end{block}
```

**Table 4.10** Official BEAMER font themes. For details and further options see [40]

| default | professionalfonts | serif |
|---|---|---|
| structurebold | structureitalicserif | structuresmallcapsserif |

```
202        \end{frame}
203
204  \end{document}
```

**Listing 4.35** Example of customized BEAMER layout, see Fig. 4.27

Some of the commands to modify element-wise the layout are summarized in Table 4.11.

## 4.2  Posters

### 4.2.1  Designing Poster based on Beamer and Beamerposter

Different specialized document classes or packages, such as a0poster [20], baposter [2], tikzposter [30], or beamerposter [10], are available to design posters. However, we will focus in this chapter on the LATEXbeamerposter package, which was developed by Philippe Dreuw and Thomas Deselaers. This package is an extension of the beamer and a0poster classes and must be, as any other LATEXpackage, initiated in the preamble of the document, see Listing 4.36. The beamerposter package allows to create posters in common standard formats such DIN-A0 or A4 in landscape or portrait orientation. Furthermore, the poster fonts can be scaled.

```
1   \documentclass[...]{beamer}
2    \usepackage[...]{beamerposter}
3    ...
4    \begin{document}
5    ...
```

**Listing 4.36** Initialization of the beamerposter package in the preamble

Users familiar with the beamer class will find it not to difficult to work with the beamerposter package. However, the official documentation comprises only a single page and the articles by de Luna [22] and Shang [32] provide a good introduction. Possible arguments of the beamerposter package are summarized in Table 4.12.

The standard font sizes of a LATEXdocument are extended and the following commands for font sizes are available for posters: \tiny, \scriptsize, \footnotesize, \small, \normalsize, \large, \Large, \LARGE, \huge, \Huge, \veryHuge, \VeryHuge, \VERYHuge. It should be noted that the option scale=x.x allows to modify these definitions by the factor x.x, see Table 4.12.

The easiest way to design a poster is to build on existing templates and examples and to introduce modifications step-by-step. We will start with a simple poster

**Fig. 4.26** Examples of different BEAMER font themes (all three versions use the presentation theme 'Berlin'): **a** default, **b** serif, and **c** structuresmallcapsserif

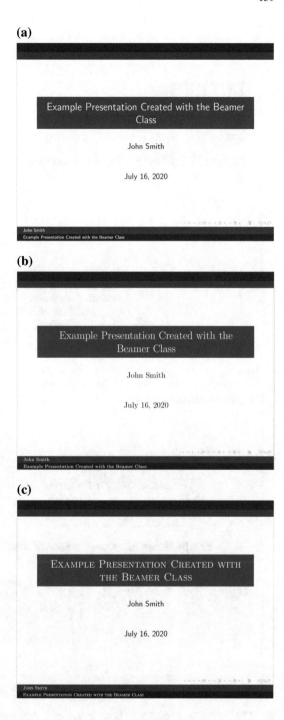

**(a)**

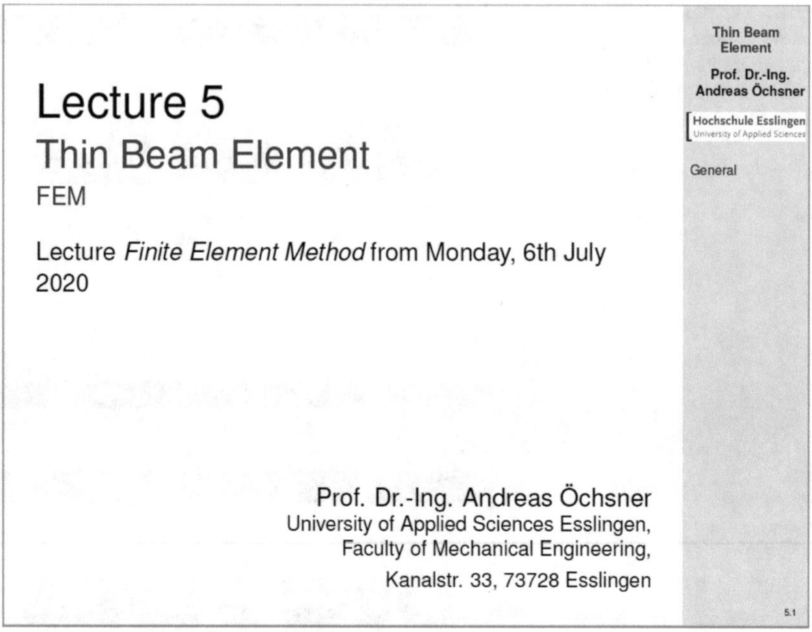

**(b)**

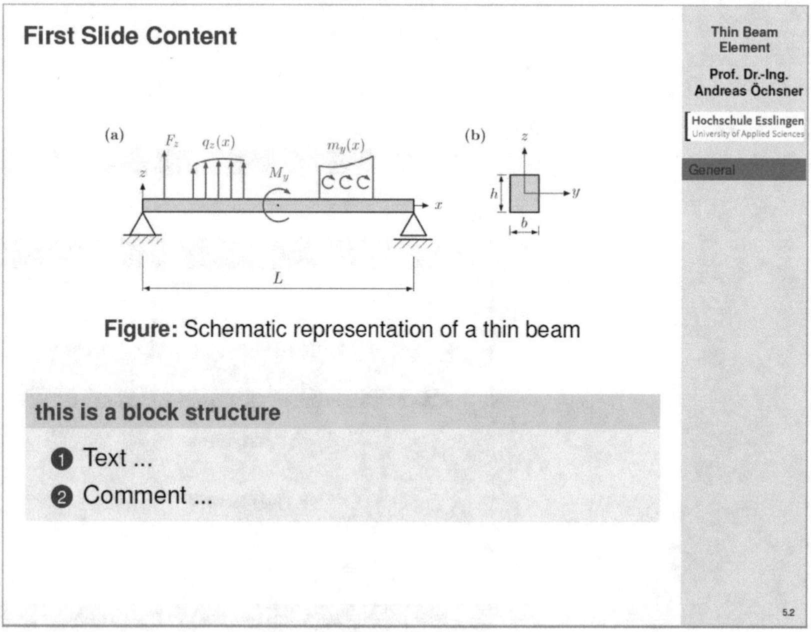

**Fig. 4.27** Example of customized BEAMER layout: **a** title page, **b** regular slide, see Listing 4.35

**Table 4.11** Some commands to customize the BEAMER layout, see [40] for more details

| Command | Comment |
|---|---|
| \usecolortheme[options]{theme} | Standard color themes are default, structure, and sidebartab or the complete color themes given in Table 4.7 |
| \usefonttheme[options]{...} | See Table 4.10 |
| \useinnertheme[options]{...} | See Table 4.9 |
| \useoutertheme[options]{...} | See Table 4.8 |
| \setbeamercolor{beamer element}{fg=blue} | Allows to define the color of the background (bg) and foreground (fg) |
| \setbeamerfont{beamer element}{size=\large,...} | In addition to size, shape (e.g. \itshape) and family (e.g. \sffamily) can be specified |
| \setbeamertemplate{element name}[predefined option]<args> | Used in different forms as in the case of predefined options (\setbeamertemplate{element name}[...] or \setbeamertemplate{element name}[...][optional argument]) or free definition (\setbeamertemplate{element name}{...}) |

**Table 4.12** Arguments of the beamerposter package

| Command | Comment |
|---|---|
| orientation=portrait | To specify the orientation. 'portrait' can be replaced by 'landscape' |
| scale=1.4 | The fonts are scaled by a factor of 1.4 |
| size=a0 | Sets the poster size to DIN-A0. Available sizes are a0, a1, a2, a3 and a4. In addition, the dimensions can be set with the options size=custom,width=x,height=y |
| debug | Adds debug information to the *.log file |

template, which can be found on the web page [3]. This template uses the official BEAMER presentation theme 'Berlin'. The first step is to introduce a header and footer by modifying the theme and calling it now 'Berlin_mod'. A source code for headers and footers can be found in the theme file written by by Philippe Dreuw and Thomas Deselaers, see the example provided on the web page [4]. Our simple example, which features a poster with a header and footer as well as the different font sizes, is given in Listing 4.37 and Fig. 4.28. The modified 'Berlin' theme file to account for the header and footer is shown in Listing 4.38.

```latex
\documentclass[final]{beamer}
\mode<presentation>{\usetheme{Berlin_mod}}
\usepackage[orientation=portrait,size=a0,scale=1.4,debug]{beamerposter}

% Data for Headline
\title[Advanced LaTeX]{Advanced \LaTeX\; in Academia}
\author[\"{O}chsner, Marco] {M.~\"{O}chsner\inst{1} and
                        \underline{A.~\"{O}chsner}\inst{2}}
\institute[short name]
            {\inst{1}%
            School of Clinical Medicine, University of Cambridge
            \and
            \inst{2}%
            Faculty of Mechanical Engineering, Esslingen\\
            University of Applied Sciences
```

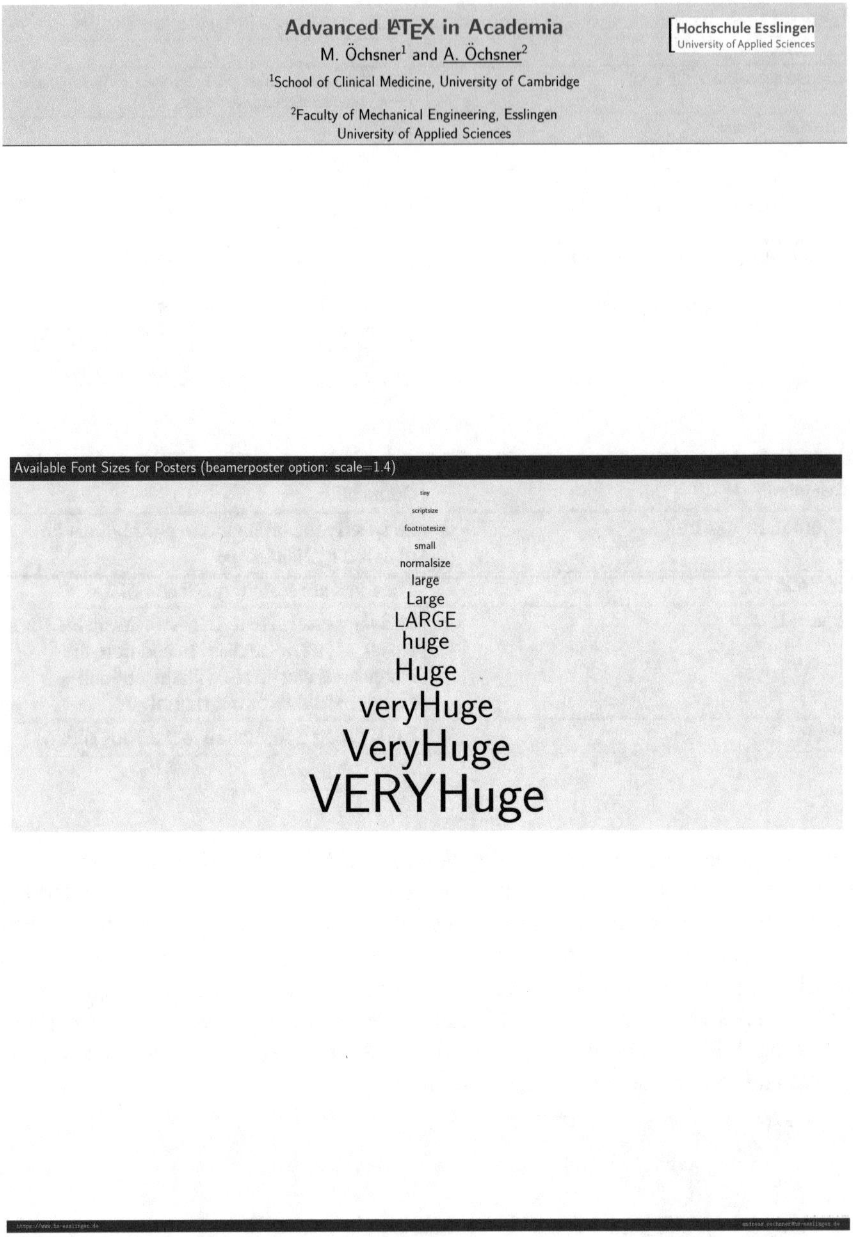

**Fig. 4.28** A simple poster with a customized header and footer, application in Listing 4.37

```
16              }
17    \date{Aug. 12th, 2020}
18    %
19
20    \begin{document}
21      \begin{frame}{}
22      \vfill
23        \begin{block}{\large Available Font Sizes for Posters
24                      (beamerposter option: scale=1.4)}
25              \centering
26              {\tiny tiny}\\[0.5ex]
27              {\scriptsize scriptsize}\\[0.5ex]
28              {\footnotesize footnotesize}\\[0.5ex]
29              {\small small}\\[0.5ex]
30              {\normalsize normalsize}\\[0.5ex]
31              {\large large}\\[0.6ex]
32              {\Large Large}\\[0.7ex]
33              {\LARGE LARGE}\\[0.7ex]
34              {\huge huge}\\[0.7ex]
35              {\Huge Huge}\\[0.8ex]
36              {\veryHuge veryHuge}\\[0.8ex]
37              {\VeryHuge VeryHuge}\\[0.8ex]
38              {\VERYHuge VERYHuge}\\[0.8ex]
39          \end{block}
40      \vfill
41      \end{frame}
42    \end{document}
```

**Listing 4.37**  A simple poster with a customized header and footer, see Fig. 4.28

```
1    \DeclareOptionBeamer{compress}{\beamer@compresstrue}
2    \ProcessOptionsBeamer
3
4    \mode<presentation>
5
6    \useoutertheme[footline=authorinstitutetitle]{miniframes}
7    \usecolortheme{whale}
8    \usecolortheme{orchid}
9    \useinnertheme{rectangles}
10
11   \setbeamerfont{block title}{size={}}
12
13   %%%%%%%%%%%%%%%%%%%%%%%%%%%%%%%%%%%%%%%%%%%%%%%%%%%%% Start Modifications
14   %%%%%%%%%%%%%%%%%%%%%%%%%%%%%%%%%%%%%%%%%%%%%%%%%%%%%%%%%%%%%%%%%%%%%%%%%%%
15
16   \setbeamercolor{headline}{fg=blue,bg=black!15}
17   \setbeamercolor{title in headline}{fg=blue}
18   \setbeamercolor{author in headline}{fg=black}
19   \setbeamercolor{institute in headline}{fg=black}
20   \setbeamercolor{separation line}{bg=blue}
21
22   %%%%%%%%%%%%%%%%%%%%%%%%%%%%%%HEADLINE
23
24   \setbeamertemplate{headline}{
25   \leavevmode
26
27   \begin{beamercolorbox}[wd=\paperwidth]{headline}
28     \begin{columns}[T]
29         \begin{column}{.01\paperwidth}
30         \end{column}
31         \begin{column}{.248\paperwidth}
32           \begin{center}
33           \vskip1cm
34         % logo on the left-hand side of the headline
35         %\includegraphics[width=.7\linewidth]{Esslingen_University_Logo.png}
36           \end{center}
37           \vskip1.5cm
```

```
38      \end{column}
39      \begin{column}{.500\paperwidth}
40        \vskip4ex
41        \centering
42        \usebeamercolor{title in headline}{\color{fg}\textbf{
43                                  \LARGE{\inserttitle}}\\[1ex]}
44        \usebeamercolor{author in headline}{\color{fg}\Large
45                                  {\insertauthor}\\[1ex]}
46        \usebeamercolor{institute in headline}{\color{fg}\large
47                                  {\insertinstitute}\\[1ex]}
48      \end{column}
49      \begin{column}{.248\paperwidth}
50        \vskip1cm
51        \begin{center}
52        % logo on the right-hand side of the headline
53        \includegraphics[width=.7\linewidth]{Esslingen_University_Logo.png}
54        \end{center}
55        \vskip1.5cm
56      \end{column}
57      \begin{column}{.01\paperwidth}
58      \end{column}
59    \end{columns}
60    \vspace*{1ex}
61    \end{beamercolorbox}
62  \begin{beamercolorbox}[wd=\paperwidth]{lower separation line head}
63    \rule{0pt}{2pt}
64  \end{beamercolorbox}
65  }
66
67  %%%%%%%%%%%%%%%%%%%%%%%%%%%FOOTLINE
68
69  \setbeamertemplate{footline}{
70  \begin{beamercolorbox}[wd=\paperwidth]{upper separation line foot}
71    \rule{0pt}{2pt}
72  \end{beamercolorbox}
73
74    \leavevmode%
75  \begin{beamercolorbox}[ht=4ex,leftskip=1cm,rightskip=1cm]{author in head/foot}%
76    \texttt{https://www.hs-esslingen.de}
77    \hfill
78    \texttt{andreas.oechsner@hs-esslingen.de}
79    \vskip1ex
80  \end{beamercolorbox}
81    \vskip0pt%
82  \begin{beamercolorbox}[wd=\paperwidth]{lower separation line foot}
83    \rule{0pt}{2pt}
84  \end{beamercolorbox}
85  }
86  %%%%%%%%%%%%%%%%%%%%%%%%%%%%%%%%%%%%%%%%%%%%%%%%%%%%%%%%%%%%%%%%%%%%%%%%%%%
87  %%%%%%%%%%%%%%%%%%%%%%%%%%%%%%%%%%%%%%%%%%%%%%%%%%%%%End Modifications
88
89  \mode
90  <all>
```

**Listing 4.38** Modified BEAMER presentation theme 'Berlin' (beamerthemeBerlin.sty): defintion of a header and footer, see Listing 4.37

```
1   \documentclass[final]{beamer}
2   \mode<presentation> {\usetheme{Berlin_mod}}
3
4   \usepackage[orientation=portrait,size=a0,scale=1.4,debug]{beamerposter}
5   \usepackage[absolute,overlay]{textpos}
6   \setlength{\TPHorizModule}{1cm}
7   \setlength{\TPVertModule}{1cm}
8   %\usepackage[colorgrid,texcoord]{eso-pic}
9
10  % Data for Headline
```

```
11  \title[Advanced LaTeX]{Advanced \LaTeX\; in Academia}
12  \author[\"{O}chsner, Marco] {M.~\"{O}chsner\inst{1} and
13                               \underline{A.~\"{O}chsner}\inst{2}}
14  \institute[short name]
15            {\inst{1}%
16            School of Clinical Medicine, University of Cambridge
17            \and
18            \inst{2}%
19            Faculty of Mechanical Engineering, Esslingen\\
20            University of Applied Sciences
21            }
22  \date{Aug. 12th, 2020}
23  %
24
25  \begin{document}
26  \begin{frame}{}
27
28  %%%% Block 1
29  \begin{textblock}{39.102}(1.45, 18.0)
30        \begin{block}{1. Introduction: The Basics}
31        \begin{itemize}
32            \item What is LaTeX?
33                \begin{itemize}
34                    \item TeX
35                    \item LaTeX
36                    \item Structure of a Document
37                \end{itemize}
38            \item Settings and Definitions
39                \begin{itemize}
40                    \item Classes
41                    \item Packages
42                    \item Cover Page
43                \end{itemize}
44        \end{itemize}
45        \end{block}
46  \end{textblock}
47
48  %%%% Block 2
49  \begin{textblock}{39.102}(43.5, 25.0)
50        \begin{block}{2. Advanced Formatting}
51        \begin{itemize}
52            \item Kerning
53            \item Font Management
54            \item Programming
55            \item Headers and Footers
56            \item Version Control
57            \item Collaborative Editing
58        \end{itemize}
59        \end{block}
60  \end{textblock}
61
62  %%%% Block 3
63  \begin{textblock}{81.102}(1.45, 50.0)
64        \begin{block}{3. Floating Objects}
65        \begin{itemize}
66            \item Figure Generation with PGFPLOTS
67            \item Figure Generation with TikZ
68                \begin{itemize}
69                    \item TikZ Matrix Library
70                    \item Flowcharts
71                    \item Remarks on Schematic Drawings
72                \end{itemize}
73            \item Tables
74        \end{itemize}
75        \end{block}
76  \end{textblock}
77
```

```
78    \end{frame}
79    \end{document}
```

**Listing 4.39**  A poster with different textblocks, see Fig. 4.29

Let us now focus on the main structure of a poster. A common procedure is to group certain elements in boxes. A first possibility is to use the LATEX package textpos together with the environment \begin{textblock}{hsize}(hpos,vpos)...\end{textblock}. This allows to create a text box of width hsize at absolute position (with its upper left corner) $x =$ hpos and $y =$ vpos, whereas the origin of the coordinate system is located at the upper left corner. The units can be set by placing the commands \setlength{\TPHorizModule}{1cm} and \setlength{\TPVertModule}{1cm} in the preamble of the document (in this case: 1 cm). A simple example with three text boxes is given in Listing 4.39 and Fig. 4.29.

It should be noted here that the option 'absolute' of the package textpos, see line 5 in Listing 4.39, ensures that all block-positioning parameters are given relative to the single origin on the page. The option 'overlay' ensures that the positioned blocks of text overlay any other page contents. The width of block 1 and 2 was obtained by the following reasoning. The A0 page of width 84 cm was divided in two columns, each having $0.49 \times$ text width. In a single column, a text block with a width of $0.95 \times$ block width should be introduced. Thus, we obtain for the width of block 1 or 2: $0.49 \times 0.95 \times 84\,\mathrm{cm} = 39.102\,\mathrm{cm}$.

To facilitate the positioning of text blocks, one can load the LATEX package esopic, see line 8 in Listing 4.39, which prints a grid over the poster, see Fig. 4.30.

An alternative approach to locate boxes with certain elements can be based on the command \begin{columns}[options]\begin{column}{widths column 1}... \begin{column}{widths column 2}....\end{columns}, which was introduced in Listing 4.6 for a classical beamer presentation. In order to ensure the horizontal alignment, a beamercolorbox and minigape construct can be used as provided in [5]. Listing 4.40 gives the same output as shown in Fig. 4.29. It should be noted that the \vspace*{...} commands were only introduced to obtain the layout of Fig. 4.29.

```
1     \documentclass[final]{beamer}
2     \mode<presentation> {\usetheme{Berlin_mod}}
3
4     \usepackage[orientation=portrait,size=a0,scale=1.4,debug]{beamerposter}
5
6     %
7     \title[Advanced LaTeX]{Advanced \LaTeX\;in Academia}
8     \author[\"{O}chsner, Marco] {M.~\"{O}chsner\inst{1} and
9                                  \underline{A.~\"{O}chsner}\inst{2}}
10    \institute[short name]
11            {\inst{1}%
12            School of Clinical Medicine, University of Cambridge
13            \and
14            \inst{2}%
15            Faculty of Mechanical Engineering, Esslingen\\
16            University of Applied Sciences
17            }
18    \date{Aug. 12th, 2020}
19
20
21    %%%%%%%%%%%% Definition beamercolorbox
22    \setbeamertemplate{block begin}{
23    \vskip.75ex
```

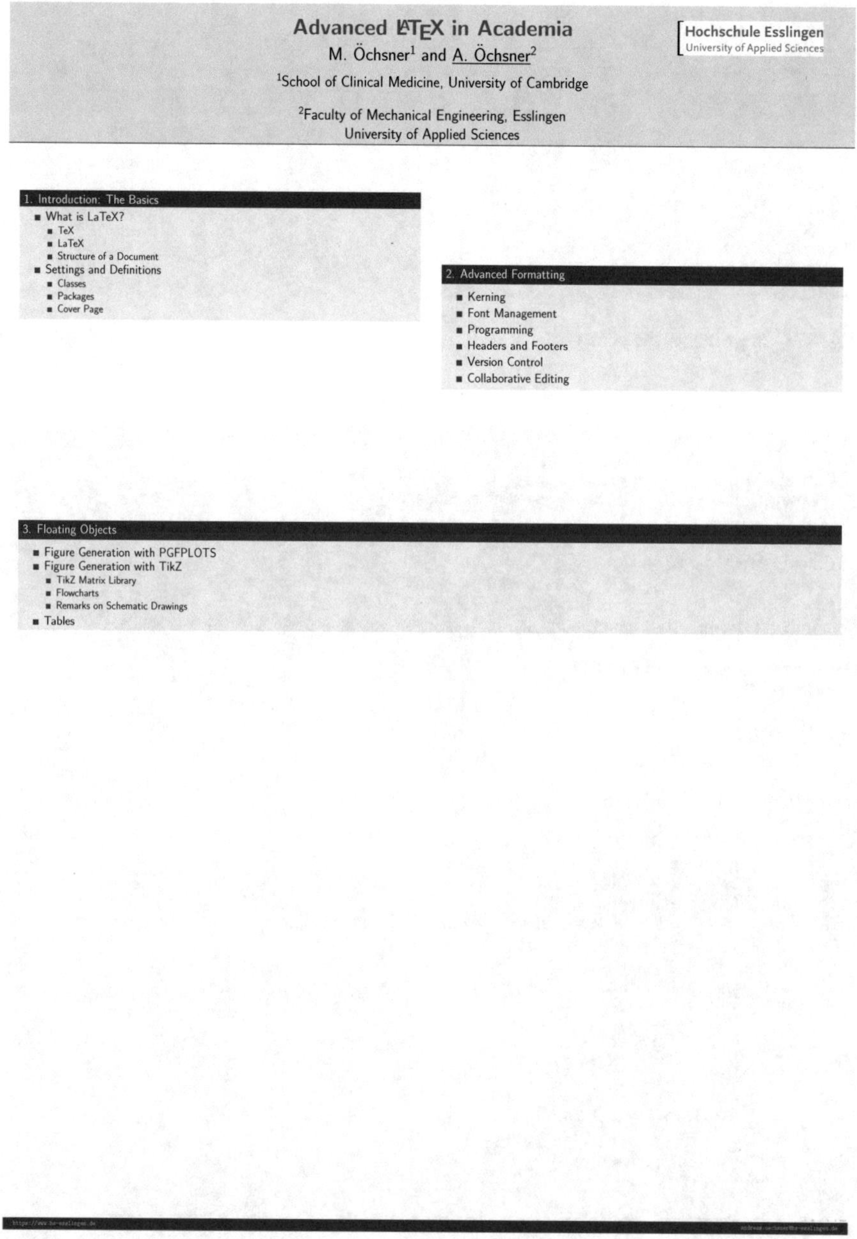

**Fig. 4.29** A poster with different textblocks, see Listing 4.39

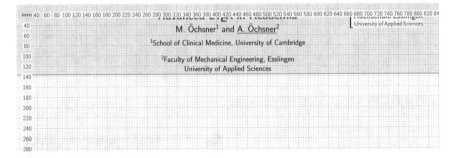

**Fig. 4.30**  A grid over the poster

```
24    \begin{beamercolorbox}[ht=3.5ex,dp=0.5ex,center,leftskip=-1em,colsep*=.75ex]
25                             {block title}%
26    \usebeamerfont*{block title}%
27    {\phantom{Gg}\insertblocktitle}% phantom because of baseline problem
28    \end{beamercolorbox}%
29    {\ifbeamercolorempty[bg]{block body}{}{\nointerlineskip\vskip-0.5pt}}%
30    \usebeamerfont{block body}%
31    \begin{beamercolorbox}[leftskip=1em,colsep*=.75ex,sep=0.5ex,vmode]
32                             {block body}%
33    \ifbeamercolorempty[bg]{block body}{\vskip-.25ex}{\vskip-.75ex}\vbox{}%
34    }
35    \setbeamertemplate{block end}{
36    \end{beamercolorbox}
37    }
38    %%%%%%%%%%%%
39
40    \begin{document}
41    \begin{frame}{}
42
43    \vspace*{-390mm}
44
45    \begin{columns}
46            \begin{column}{.49\textwidth}
47            \begin{beamercolorbox}[center,wd=\textwidth]{postercolumn}
48            \begin{minipage}[T]{.95\textwidth}
49            %%%% 1
50            \begin{block}{1. Introduction: The Basics}
51            \begin{itemize}
52            \item What is LaTeX?
53            \begin{itemize}
54            \item TeX
55            \item LaTeX
56            \item Structure of a Document
57            \end{itemize}
58            \item Settings and Definitions
59            \begin{itemize}
60            \item Classes
61            \item Packages
62            \item Cover Page
63            \end{itemize}
64            \end{itemize}
65            \end{block}
66            \end{minipage}
67            \end{beamercolorbox}
68            \end{column}
69            \begin{column}{.49\textwidth}
70            \vspace*{13cm}
```

```
71    \begin{beamercolorbox}[center,wd=\textwidth]{postercolumn}
72    \begin{minipage}[T]{.95\textwidth}
73    %&&&2
74    \begin{block}{2. Advanced Formatting}
75    \begin{itemize}
76    \item Kerning
77    \item Font Management
78    \item Programming
79    \item Headers and Footers
80    \item Version Control
81    \item Collaborative Editing
82    \end{itemize}
83    \end{block}
84    \end{minipage}
85    \end{beamercolorbox}
86    \end{column}
87 \end{columns}
88
89 \vspace*{120mm}
90
91 \begin{columns}
92    \begin{column}{0.98\textwidth}
93    \begin{beamercolorbox}[center,wd=\textwidth]{postercolumn}
94    \begin{minipage}[T]{0.99\textwidth}
95    %&&&3
96    \begin{block}{3. Floating Objects}
97    \begin{itemize}
98    \item Figure Generation with PGFPLOTS
99    \item Figure Generation with TikZ
100   \begin{itemize}
101   \item TikZ Matrix Library
102   \item Flowcharts
103   \item Remarks on Schematic Drawings
104   \end{itemize}
105   \item Tables
106   \end{itemize}
107   \end{block}
108   \end{minipage}
109   \end{beamercolorbox}
110   \end{column}
111 \end{columns}
112
113 \end{frame}
114 \end{document}
```

**Listing 4.40** A poster structure based on the columns command, see Fig. 4.29

## 4.2.2  Decomposing a Poster

Large scale posters, e.g. in A0 format, can be decomposed to A4 pages and thus, printed on a common A4 printer. The user may glue the single A4 pages together and proofread or present the A0 poster. This task can be done by incorporating the LATEX package pdfpages and compiling the document with pdflatex. To do so, the A0 PDF poster should be named, for example, A0poster.pdf and located in the same directory as the file in Listing 4.41. Compiling the file of Listing 4.41 gives a PDF file with 16 single A4 pages, which compose the entire A0 poster.

```
1   \documentclass[a4paper]{article}
2
3   \usepackage{pdfpages}
4   \begin{document}
5     % top row, left to right
6     \includepdf[viewport= 0 2526 596 3368]{A0poster.pdf}
7     \includepdf[viewport= 595 2526 1192 3368]{A0poster.pdf}
8     \includepdf[viewport=1190 2526 1788 3368]{A0poster.pdf}
9     \includepdf[viewport=1785 2526 2384 3368]{A0poster.pdf}
10    % 2nd row, left to right
11    \includepdf[viewport= 0 1684 596 2526]{A0poster.pdf}
12    \includepdf[viewport= 595 1684 1192 2526]{A0poster.pdf}
13    \includepdf[viewport=1190 1684 1788 2526]{A0poster.pdf}
14    \includepdf[viewport=1785 1684 2384 2526]{A0poster.pdf}
15    % 3rd row, left to right
16    \includepdf[viewport= 0 842 596 1684]{A0poster.pdf}
17    \includepdf[viewport= 595 842 1192 1684]{A0poster.pdf}
18    \includepdf[viewport=1190 842 1788 1684]{A0poster.pdf}
19    \includepdf[viewport=1785 842 2384 1684]{A0poster.pdf}
20    % bottom row, left to right
21    \includepdf[viewport= 0 0 596 842]{A0poster.pdf}
22    \includepdf[viewport= 595 0 1192 842]{A0poster.pdf}
23    \includepdf[viewport=1190 0 1788 842]{A0poster.pdf}
24    \includepdf[viewport=1785 0 2384 842]{A0poster.pdf}
25  \end{document}
```

**Listing 4.41**  Decomposing of a A0 poster into A4 pages. Adapted from [22]

# Chapter 5
# Exams, Tests and Quizzes

**Abstract** This chapter introduces the exam document class which allows to generate exams, test, and quizzes. Different ways of defining questions and ways of providing the answers are introduced. The package also allows to display on request the solutions to the questions and to automatically create grading and point tables.

## 5.1 The exam Document Class

Different specialized document classes, such as exam [18], eqexam [34], exercise [29], or exsheets [23], are available. However, we will focus in this chapter on the exam document class, which is developed by Philip Hirschhorn from the Department of Mathematics at Wellesley College, and comes with more than 130 pages of documentation, see [18] for details. The document class must be, as all other classes, initiated in the preamble of the document, see Listing 5.1.

```
1  \documentclass[...]{exam}
2  ...
3  \begin{document}
4  ...
```

**Listing 5.1** Initialization of the exam class in the preamble

Possible arguments are, for example, 12pt to specify the font size, a4paper to specify the page size, addpoints to add up the points for grading and point tables (see Sect. 5.5), and answers to print solutions into the exam (see Sect. 5.4).

## 5.2 Headers and Footers

To indicate if headers or footers should be used or not, one can define with the \pagestyle{...} command different modifications, see Table 5.1. Important is that this command must be given before the \begin{document} command. It should be noted here that the default page style is headandfoot whereas the default header is empty and the default footer displays the page number in the center.

M. Öchsner and A. Öchsner, *Advanced LaTeX in Academia*,
https://doi.org/10.1007/978-3-030-88956-2_5

**Table 5.1**  Commands for initialization of headers and footers

| Command | Comment |
| --- | --- |
| \pagestyle{headandfoot} | Header and footer can be defined |
| \pagestyle{head} | Only header can be defined |
| \pagestyle{foot} | Only footer can be defined |
| \pagestyle{empty} | No header and no footer |
| \thispagestyle{...} | To change the style on a single page |

Headers and footers are generally structured in three parts, i.e. a left-hand, a centered, and a right-hand part. These parts can be independently defined based on two approaches where it is additionally distinguished if the first page is different from the following ones or not. Let us focus on the header and only collect the corresponding commands for the footer, see Table 5.2. The command **\header{left-hand text}{centered text}{right-hand text}** allows to define these three parts uniformly for all pages. If the header, for example, for a cover sheet must be different to all following pages, the commands **\firstpageheader{...}{...}{...}** and **\runningheader{...}{...}{...}** allow to define different headers. Alternatively, the commands **\lhead[]{...}**, **\chead[]{...}**, and **\rhead[]{...}** could be used to define separately the three parts of the header where the optional text in the square brackets would be only applied to the first page and the pages after the first are defined by the text in the {...} structure. The corresponding commands for the footer are collected in Table 5.2. Listing 5.2 and Figs. 5.1 and 5.2 illustrate a simple application of the header and footer commands.

```
1   \pagestyle{headandfoot}
2
3   \lhead{\bfseries\large ADV. FEM\\Professor \"{O}chsner}
4   \chead{}
5   \rhead{\bfseries\large Final Exam\\25.01.2020}
6
7   \footer{}{Page \thepage\ of \numpages}{}
8
9
10  \begin{document}
11  ...
```

**Listing 5.2**  Environment for a simple header and footer structure

An advanced header with different versions for the first and following pages is illustrated in Listing 5.3 and Figs. 5.3 and 5.4. To create the table style header, a table is created in a minipage and lrbox and this named construct is called in the header commands, i.e. **firstpageheader** and **runningheader**.

```
1   ...
2   \pagestyle{headandfoot}
3   \extraheadheight{35.0mm}
4
5   \usepackage{tabularx}
6   \usepackage{graphicx}
7
8   \newlength{\headerwidth}
9   \setlength{\headerwidth}{\textwidth}
```

**Table 5.2** Commands for defining headers and footers

| Command | Comment |
|---|---|
| Header | |
| \header{...}{...}{...} | Header with left-hand, centered, and right-hand text, uniform for all pages |
| \lhead{...} | Left-hand text for header, uniform for all pages |
| \chead{...} | Centered text for header, uniform for all pages |
| \rhead{...} | Right-hand text for header, uniform for all pages |
| \firstpageheader{...}{...}{...} | Header of first page with left-hand, centered, and right-hand text |
| \runningheader{...}{...}{...} | Header of pages after the first one with left-hand, centered, and right-hand text |
| \lhead[...]{...} | Left-hand text for header, on first page as [...] and on following pages as {...} |
| \chead[...]{...} | Centered text for header, on first page as [...] and on following pages as {...} |
| \rhead[...]{...} | Right-hand text for header, on first page as [...] and on following pages as {...} |
| Footer | |
| \footer{...}{...}{...} | Footer with left-hand, centered, and right-hand text, uniform for all pages |
| \lfoot{...} | Left-hand text for footer, uniform for all pages |
| \cfoot{...} | Centered text for footer, uniform for all pages |
| \rfoot{...} | Right-hand text for footer, uniform for all pages |
| \firstpagefooter{...}{...}{...} | Footer of first page with left-hand, centered, and right-hand text |
| \runningfooter{...}{...}{...} | Footer of pages after the first one with left-hand, centered, and right-hand text |
| \lfoot[...]{...} | Left-hand text for footer, on first page as [...] and on following pages as {...} |
| \cfoot[...]{...} | Centered text for footer, on first page as [...] and on following pages as {...} |
| \rfoot[...]{...} | Right-hand text for footer, on first page as [...] and on following pages as {...} |

**ADV. FEM**
**Professor Öchsner**

**Final Exam**
**25.01.2020**

**Fig. 5.1** A simple header for all pages, see Listing 5.2

**Fig. 5.2** A simple footer for all pages, see Listing 5.2

Page 1 of 2

## Esslingen University

⌈ Hochschule Esslingen
 University of Applied Sciences

| Winter Semester 2019/2020 | Professor Dr.-Ing. A. Öchsner, D.Sc. |
|---|---|
| Faculty of Mechanical Engineering | Semester: DDM1 |
| Course: Light W. D. / Ad. Fin. El. M. | Exam Code: 1221005 |
| Date: 25.01.2020 | Time Duration: 120 minutes |
| Remarks: closed book exam; all lecture slides allowed | |

**Name:**                                                    **Matr. Number:**

**Fig. 5.3** An advanced header for the first page, see Listing 5.3

| Winter Semester 2019/2020 | Professor Dr.-Ing. A. Öchsner, D.Sc. |
|---|---|
| Faculty of Mechanical Engineering | Semester: DDM1 |
| Course: Light W. D. / Ad. Fin. El. M. | Exam Code: 1221005 |
| Date: 25.01.2020 | Time Duration: 120 minutes |
| Remarks: closed book exam; all lecture slides allowed | |

**Fig. 5.4** An advanced header for all pages except the first, see Listing 5.3

```
10
11   % Header first page
12
13   \newsavebox{\myheader}
14    \begin{lrbox}{\myheader}%
15    \begin{minipage}[b]{\headerwidth}
16    \renewcommand{\arraystretch}{1.29}%
17    \begin{tabularx}{\headerwidth}{|X|X|}
18    \multicolumn{1}{|l|}{\bfseries \Large Esslingen University }&
19    \multicolumn{1}{r|}{\includegraphics[scale=0.2]{Logo.png}}\\
20    \hline
21    Winter Semester 2019/2020          & Professor Dr.-Ing. A. \"{O}chsner, D.Sc.\\
22    \hline
23    Faculty of Mechanical Engineering & Semester: DDM1 \\
24    \hline
25    Course: Light W.D. / Ad.Fin.El.M. & Exam Code: 1221005\\
26    \hline
27    Date: 25.01.2020               & Time Duration: 120 minutes\\
28    \hline
29    \multicolumn{2}{|l|}{Remarks: closed book exam; all lecture slides allowed}\\
30    \hline
31    \end{tabularx}
32    \vspace*{1mm}
33    \par
34    \bfseries
35    Name:  \hspace*{80mm}
36    Matr. Number:          \hspace*{\fill}
37    \end{minipage}
38    \end{lrbox}
39
40   % Header for all pages except the first
41
42   \newsavebox{\myheaderrun}
43    \begin{lrbox}{\myheaderrun}%
```

```
44    \begin{minipage}[b]{\headerwidth}
45    \renewcommand{\arraystretch}{1.29}%
46    \begin{tabularx}{\headerwidth}{|X|X|}
47    \hline
48    Winter Semester 2019/2020              & Professor Dr.-Ing. A. \"{O}chsner, D.Sc.\\
49    \hline
50    Faculty of Mechanical Engineering & Semester: DDM1 \\
51    \hline
52    Course:  Light W.D.  /  Ad.Fin.El.M. & Exam Code: 1221005\\
53    \hline
54    Date: 25.01.2020                       & Time Duration: 120 minutes\\
55    \hline
56    \multicolumn{2}{|l|}{Remarks: closed book exam; all lecture slides allowed}\\
57    \hline
58    \end{tabularx}
59    \end{minipage}
60    \end{lrbox}
61
62    \firstpageheader{}{\usebox{\myheader}}{}
63    \runningheader{}{\usebox{\myheaderrun}}{}
64
65    \footer{}{Page \thepage\ of \numpages}{}
66
67    \begin{document}
68    ...
```

**Listing 5.3**  Environment for advanced header with different layout for the first and following pages

The command \extraheadheight{35.0mm} (see Listing 5.3, line 3) allows introducing some additional space between the header and the top border of your page. If a different value is required for the first page and the pages after the first, the optional argument allows this differentiation: \extraheadheight[45.0mm]{35.0mm}.

## 5.3  Questions and Points

Questions are embedded in the questions environment \begin{questions} ... \end{questions} and each question begins with a \question command. A number inside square brackets allows to display the number of points[1] assigned to this question, see Listing 5.4 for the source code and Fig. 5.5 for the output.

```
1    \begin{questions}
2        \question[2]
3            What is the Young's modulus of steel?
4        \question
5            What is the Young's modulus of aluminum?
6    \end{questions}
```

**Listing 5.4**  Environment for a simple question structure

To change the default style of the first question line, i.e., the question number to appear at the left-hand margin and for the text of the question to begin on that line, the command \qformat{...} can be used. Listing 5.5 and Fig. 5.6 display how the word 'Question' can be introduced and the corresponding question text to begin on the following line.

---

[1] The command \pointpoints{Punkt}{Punkte} stated before \begin{document} allows to replace the words 'point' and 'points', for example, in another language.

1. (2 points)  What is the Young's modulus of steel?

2. What is the Young's modulus of aluminum?

**Fig. 5.5**  A simple question structure, see Listing 5.4

**Question 1**   (2 points)
What is the Young's modulus of steel?

**Question 2**   (2 points)
What is the Young's modulus of aluminum?

**Fig. 5.6**  A simple question structure with the wording 'Question', see Listing 5.5

```
\begin{questions}
\qformat\textbfQuestion \thequestion\quad (\thepoints)\hfill
\question[2]
    What is the Young's modulus of steel?
\question[2]
    What is the Young's modulus of aluminum?
\end{questions}
```

**Listing 5.5**  Environment for a simple question structure with the wording 'Question'

Under certain circumstances it might be required to change the standard wording (e.g. 'Question' and 'points') as shown Fig. 5.6. This might be the case, for example, if the language used must be changed. Listing 5.6 indicates the required commands for changing the standard wording (see Fig. 5.7 for the output).

```
\begin{questions}
\pointpoints{Punkt}{Punkte}
\qformat{\textbf{Aufgabe \thequestion}\quad (\thepoints)\hfill}
\question[2]
    ... ?
\question[1]
    ... ?
\end{questions}
```

**Listing 5.6**  Environment for a simple question structure with different wording for 'Question' and 'points'

To adjust the question numbering, the command \setcounter{question}{...} can be used in the question environment, see Listing 5.7 for the source code and Fig. 5.8 for the output.

```
\begin{questions}
\setcounter{question}{24}
  \question[2]
        What is the Young's modulus of steel?
  \question[2]
        What is the Young's modulus of aluminum?
\end{questions}
```

**Listing 5.7**  Environment for a simple question structure with renumbered question numbers

**Aufgabe 1** (2 Punkte)
... ?

**Aufgabe 2** (1 Punkt)
... ?

**Fig. 5.7** A simple question structure with different wording for 'Question' and 'points', see Listing 5.6

25. (2 points) What is the Young's modulus of steel?

26. (2 points) What is the Young's modulus of aluminum?

**Fig. 5.8** A simple question structure with renumbered question numbers, see Listing 5.7

1. State the Young's moduli of the following engineering materials:
   (a) (2 points) steel,
   (b) (2 points) aluminum.

**Fig. 5.9** A subdivided question structure, see Listing 5.8

To further structure or subdivide the questions environment, the commands \part, \subpart, and \subsubpart are available. The commands can be used immediately after the initiation of the questions environment or later after some initial text, see Listing 5.8 for the source code and Fig. 5.9 for the output.

```
\begin{questions}
\question
  State the Young's moduli of the following engineering materials:
\begin{parts}
    \part[2] steel,
    \part[2] aluminum.
\end{parts}
\end{questions}
```

**Listing 5.8** Environment for a subdivided question structure

In case that decimal half points (e.g. 1.5) should be indicated, one may define a new command in the preamble of the document as indicated in Listing 5.9.

```
\documentclass[...]{exam}
\renewcommand*\half{.5}
...
\begin{document}
...
\begin{questions}
\question
  State the Young's moduli of the following engineering materials:
\begin{parts}
    \textbackslash part[2] steel,
    \textbackslash part[1\half] aluminum.
\end{parts}
\end{questions}
```

**Listing 5.9** Defintion of decimal half points

1. (2 points)  What is the Young's modulus of steel?




2. (2 points)  What is the Young's modulus of aluminum?

**Fig. 5.10**  A question structure with blanc space for the answers, see Listing 5.10


To further structure the question environment, different types of spaces for the answers can be introduced. A simple space can be introduced with the **\vspace{length}** command (see Listing 5.10 for the source code and Fig. 5.10 for the output), whereas the **\vspace*{\stretch{1}}** command allows to stretch the free space till the end of the page or to equally distribute the space in case of several questions. Furthermore, a different number than '1' in the **\vspace*{\stretch{1}}** command allows to unevenly distribute the allocated space.

```
1  \begin{questions}
2    \question[2]
3      What is the Young's modulus of steel?
4      \vspace{20mm}
5    \question[2]
6      What is the Young's modulus of aluminum?
7      \vspace{20mm}
8  \end{questions}
```

**Listing 5.10**  Environment for a question structure with blanc space for the answers

To better indicate the available space for an answer, a framed box can be introduced with the command **\makeemptybox{length}**, see Listing 5.11 for the source code and Fig. 5.11 for the output. The command **\makeemptybox{\stretch{1}}** allows again to stretch a box till the end of the page or equally distribute the space in case of several boxes.

```
1  \begin{questions}
2    \question[2]
3      What is the Young's modulus of steel?
4      \makeemptybox{20mm}
5    \question[2]
6      What is the Young's modulus of aluminum?
7      \makeemptybox{20mm}
8  \end{questions}
```

**Listing 5.11**  Environment for a question structure with empty boxes for the answers

Alternatively, the available space for an answer can be indicated by lines (command **\fillwithlines{length}**),[2] see Listing 5.12 for the source code and Fig. 5.12 for the output, dotted lines (command **\fillwithdottedlines{length}**), or a grid (command **\fillwithgridlength**),[3] see Listing 5.13 for the source code and Fig. 5.13 for the output.

---

[2] The command \linefillheight allows to adjust the line spacing. This command can be used in a question environment for single use or before \begin{document} for a global definition.

[3] The commands \setlength{\gridsize}{length} and \setlength{\gridlinewidth}{thickness} allow to adjust the grid spacing and the thickness of the grid lines.

1. (2 points) What is the Young's modulus of steel?

<br><br><br><br><br><br>

2. (2 points) What is the Young's modulus of aluminum?

<br><br><br><br><br><br>

**Fig. 5.11** A question structure with empty boxes for the answers, see Listing 5.11

1. (2 points) What is the Young's modulus of steel?

<br><br><br><br>

2. (2 points) What is the Young's modulus of aluminum?

<br><br><br><br>

**Fig. 5.12** A question structure with lines for the answers, see Listing 5.12

```
\begin{questions}
  \question[2]
    What is the Young's modulus of steel?
    \fillwithlines{20mm}
  \question[2]
    \setlength{\linefillheight}{2mm}
    What is the Young's modulus of aluminum?
    \fillwithlines{20mm}
\end{questions}
```

**Listing 5.12** Environment for a question structure with lines for the answers

```
\begin{questions}
  \question[2]
    What is the Young's modulus of steel?
    \fillwithgrid{20mm}
  \question[2]
    \setlength{\gridsize}{7.5mm}
    \setlength{\gridlinewidth}{0.3pt}
    What is the Young's modulus of aluminum?
    \fillwithgrid{30mm}
\end{questions}
```

**Listing 5.13** Environment for a question structure with a grid for the answers

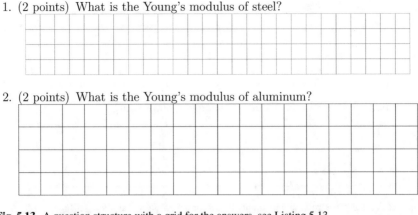

1. (2 points) What is the Young's modulus of steel?

2. (2 points) What is the Young's modulus of aluminum?

**Fig. 5.13** A question structure with a grid for the answers, see Listing 5.13

1. (2 points) What is the Young's modulus of steel?

                                         1. _____

2. (2 points) What is the Young's modulus of aluminum?

                                         2. _____

**Fig. 5.14** A question structure with a short space for the answers, see Listing 5.14

Some space for short answers can be realized with the command \answerline[answer]. The optional argument in the brackets allows to write the answer and this can be displayed on request, see Sect. 5.4. The following Listing 5.14 and Fig. 5.14 show a simple application of this command.

```
1  \begin{questions}
2      \question[2]
3          What is the Young's modulus of steel?
4          \answerline[210000 MPa]
5      \question[2]
6          What is the Young's modulus of aluminum?
7          \answerline[70000 MPa]
8  \end{questions}
```

**Listing 5.14** Environment for a question structure with a short space for the answers

Incorporation of figures in an question environment are possible via the classical \includegraphics command, see Listing 5.15 for the source code and Fig. 5.15 for the output.

```
1  \begin{questions}
2      \question[3]
3          Sketch a typical stress strain diagram of a ductile steel!
4          \begin{center}
5          \includegraphics[scale=1.2]{axis.eps}
6          \end{center}
7  \end{questions}
```

**Listing 5.15** Environment for a question structure with an included figure in the question environment

    1. (3 points) Sketch a typical stress strain diagram of a ductile steel!

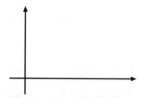

**Fig. 5.15** A question structure with an included figure in the question environment, see Listing 5.15

    1. (3 points) Sketch a typical stress strain diagram of a ductile steel!

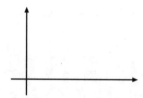

Figure 1: Schematic stress strain diagram

**Fig. 5.16** A question structure with an included figure with number and caption in the question environment, see Listing 5.16

If a figure number and caption are required, one may place the \includegraphics command within an \begin{figure} ... \end{figure} environment, see Listing 5.16 for the source code and Fig. 5.16 for the output.

```
1   \begin{questions}
2     \question[3]
3       Sketch a typical stress strain diagram of a ductile steel!
4       \begin{figure}[h]
5         \centering
6         \includegraphics[scale=1.2]{axis.eps}
7         \caption{Schematic stress strain diagram}
8       \end{figure}
9   \end{questions}
```

**Listing 5.16** Environment for a question structure with an included figure with number and caption in the question environment

Four different methods for structuring multiple choice questions are available in the exam class. The first option, which is embedded in a **choices** environment, provides upper case letters to distinguish the different options for the answers, see Listing 5.17 for the source code and Fig. 5.17 for the output. Thus, the correct answers could be circled on the answering sheet. The \CorrectChoice command allows to indicate the correct answer and this can be displayed on request, see Sect. 5.4.

1. What is the Young's modulus of steel?

   A. 70000 MPa

   B. 210000 MPa

   C. 110000 MPa

**Fig. 5.17** Multiple choice question structure with upper case letters, see Listing 5.17

1. What is the Young's modulus of steel?

   ○ 70000 MPa

   ○ 210000 MPa

   ○ 110000 MPa

**Fig. 5.18** Environment for a multiple choice question structure with checkboxes, see Listing 5.18

```
1  \begin{questions}
2      \question What is the Young's modulus of steel?
3      \begin{choices}
4          \choice 70000 MPa
5          \CorrectChoice 210000 MPa
6          \choice 110000 MPa
7      \end{choices}
8  \end{questions}
```

**Listing 5.17** Environment for a multiple choice question structure with upper case letters

The second option for structuring multiple choice questions is to replace the upper case letters by checkboxes. The **checkboxes** environment is illustrated in Listing 5.18 and Fig. 5.18.

```
1  \begin{questions}
2      \question What is the Young's modulus of steel?
3      \begin{checkboxes}
4          \choice 70000 MPa
5          \CorrectChoice 210000 MPa
6          \choice 110000 MPa
7      \end{checkboxes}
8  \end{questions}
```

**Listing 5.18** Environment for a multiple choice question structure with checkboxes

The third option for structuring multiple choice questions is to arrange the checkboxes in a single paragraph. This **oneparcheckboxes** environment is illustrated in Listing 5.19 and Fig. 5.19.

```
1  \begin{questions}
2      \question What is the Young's modulus of steel?
3      \begin{oneparcheckboxes}
4          \choice 70000 MPa
5          \CorrectChoice 210000 MPa
6          \choice 110000 MPa
7      \end{oneparcheckboxes}
8  \end{questions}
```

**Listing 5.19** Environment for a multipe choice question structure with checkboxes and listing in a single paragraph

1. What is the Young's modulus of steel?  ○ 70000  ○ 210000  ○ 110000 MPa

1. _____ is the Young's modulus of steel

**Fig. 5.19** Multiple choice question structure with checkboxes and listing in a single paragraph, see Listing 5.19

It should be noted here that the multiple choice question structure with upper case letters (see Listing 5.17) can be also printed in a single paragraph by replacing the **choices** environment with an **oneparchoices** environment (fourth option for a multiple choice question).

## 5.4 Solutions

The exam class provides comprehensive possibilities to display the correct answers on request. It is important to use the class option **answers** if the included solutions should be displayed, see Listing 5.20. The additional class option **cancelspace** causes the provided blank space to be deleted, and instead only the solution is displayed. The solution itself can be written in the \begin{solution}[size]...\end{solution} environment which is included in the questions environment. The optional parameter **size** allows to define a blanc space equivalent to the \vspace{size} command as shown in Listing 5.10, i.e. for the version without displayed solution. Other ways to indicate the space for the solution, e.g. an empty box or lines space, can be achieved by replacing the solution environment by other options, see Table 5.3 for details. Listing 5.20 and Fig. 5.20 illustrate the inclusion of solutions in a question structure with blanc space for the answers. The default setting of this listing displays the solution in a box whose width equals that of the text of the current question.

```
1   % version without solutions
2   % \documentclass[a4paper,12pt]{exam}
3   % version with solutions
4   \documentclass[a4paper,12pt,answers,cancelspace]{exam}
5   ...
6   \begin{questions}
7     \question[2]
8       What is the Young's modulus of steel?
9   %
10    \begin{solution}[20mm]
11      210000 MPa
12    \end{solution}
13    \question[2]
14      What is the Young's modulus of aluminum?
15  %
16    \begin{solution}[20mm]
17      70000 MPa
18    \end{solution}
19  \end{questions}
```

**Listing 5.20** Inclusion of solutions in a question structure with space for the answers

To alternatively print the solution on a shaded background in light gray color (default setting), the command **\shadedsolutions** can be used as shown in Listing 5.21. In addition, the inclusion of the usepackage color is required, see also Fig. 5.21 for the output.

```
1   % version without solutions
2   % \documentclass[a4paper,12pt]{exam}
3   % version with solutions
4   \documentclass[a4paper,12pt,answers,cancelspace]{exam}
5   \usepackage{color}
6   ...
7   \shadedsolutions
8   \begin{questions}
9       \question[2]
10          What is the Young's modulus of steel?
11  %
12      \begin\solution}[20mm]
13          210000 MPa
14      \end{solution}
15      \question[2]
16          What is the Young's modulus of aluminum?
17  %
18      \begin{solution}[20mm]
19          70000 MPa
20      \end{solution}
21  \end{questions}
```

**Listing 5.21** Inclusions of solutions on a shaded background in a question structure with space for the answers

**Table 5.3** Commands for the definition of the solution space style

| Command without a solution environment | Command with a solution environment | Type of solution space |
|---|---|---|
| \vspace{size} | \begin{solution}[space]... | Blank space as in Fig. 5.10 |
| \makeemptybox{size} | \begin{solutionorbox}[space]... | Empty box as in Fig. 5.11 |
| \fillwithlines{size} | \begin{solutionorlines}[space]... | Lined space as in Fig. 5.12 |
| \fillwithdottedlines{size} | \begin{solutionordottedlines}[space]... | Dotted lined space |
| \fillwithgrid{size} | \begin{solutionorgrid}[space]... | Space filled with a grid as in Fig. 5.13 |

1. (2 points) What is the Young's modulus of steel?

> **Solution:** 210000 MPa

2. (2 points) What is the Young's modulus of aluminum?

> **Solution:** 70000 MPa

**Fig. 5.20** Inclusion of solutions in a question structure with space for the answers, see Listing 5.20

1. (2 points) What is the Young's modulus of steel?

> **Solution:** 210000 MPa

2. (2 points) What is the Young's modulus of aluminum?

> **Solution:** 70000 MPa

**Fig. 5.21** Inclusions of solutions on a shaded background in a question structure with space for the answers, see Listing 5.21

1. What is the Young's modulus of steel?

A. 70000 MPa

**B. 210000 MPa**

C. 110000 MPa

**Fig. 5.22** Displaying the solution of a multiple choice question with upper case letters, see Listing 5.17 and Fig. 5.17

1. What is the Young's modulus of steel?

○ 70000 MPa

√ **210000 MPa**

○ 110000 MPa

**Fig. 5.23** Displaying the solution of a multiple choice question with checkboxes, see Listing 5.18 and Fig. 5.18

It should be noted here that the light gray background can be easily changed with the following command, using the RGB color model[4]

\definecolor{SolutionColor}{rgb}{0.95 ,0.44 ,0.13},

or in the CMYK color model

\definecolor{SolutionColor}{cmyk}{0.0,0.7,1.0,0.0}.

If the solutions should be printed without any framing or shading, the command \unframedsolutions switches these options off.

The display of solutions in the case of multiple choice questions is much easier than in the case of provided blank space for the answer. The command \CorrectChoice instead of \Choice allows for indicating the correct answer (see Listings 5.17 and 5.18) and Figs. 5.22 and 5.23 illustrate how the correct answer is displayed if the class option answers is used.

---

[4] See Sect. A for further information on color definitions.

## 5.5   Grading and Point Tables

Grading and point tables can be easily generated and allow to display the points
per question or page and the total possible number of points. In addition, a grading
table provides space to manually introduce the obtained points for each question. To
automatically add the points together, the option addpoints must be used as a class
option. A simple example for the definition of a grading and point table is shown in
Listings 5.22 and 5.23 and the corresponding outputs in Figs. 5.24 and 5.25.

```
1  \documentclass[a4pape,12pt,addpoints]{exam}
2  ...
3  \begin{questions}
4    \question[2]
5      Explain the expression \emph{isotropic} material.
6    \question[2]
7      Explain the expression \emph{homogeneous} material.
8    \question
9      State the Young's moduli of the following engineering materials:
10   \begin{parts}
11     \part[2] steel,
12     \part[2] aluminum.
13   \end{parts}
14 \end{questions}
15
16 \begin{center}
17   \gradetable[v][questions]
18 \end{center}
```

**Listing 5.22**   Definition of a simple grading table

1. (2 points) Explain the expression *isotropic* material.

2. (2 points) Explain the expression *homogeneous* material.

3. State the Young's moduli of the following engineering materials:

   (a) (2 points) steel,

   (b) (2 points) aluminum.

| Question | Points | Score |
|:--------:|:------:|:-----:|
| 1 | 2 | |
| 2 | 2 | |
| 3 | 4 | |
| Total: | 8 | |

**Fig. 5.24**   Definition of a simple grading table, see Listing 5.22

1. (2 points) Explain the expression *isotropic* material.

2. (2 points) Explain the expression *homogeneous* material.

3. State the Young's moduli of the following engineering materials:

    (a) (2 points) steel,

    (b) (2 points) aluminum.

| Question | Points |
|:--------:|:------:|
| 1 | 2 |
| 2 | 2 |
| 3 | 4 |
| Total: | 8 |

**Fig. 5.25** Definition of a simple point table, see Listing 5.23

```
1   \documentclass[a4pape,12pt,addpoints]{exam}
2   ...
3   \begin{questions}
4       \question[2]
5           Explain the expression \emph{isotropic} material.
6       \question[2]
7           Explain the expression \emph{homogeneous} material.
8       \question
9           State the Young's moduli of the following engineering materials:
10      \begin{parts}
11          \part[2] steel,
12          \part[2] aluminum.
13      \end{parts}
14  \end{questions}
15
16  \begin{center}
17      \pointtable[v][questions]
18  \end{center}
```

**Listing 5.23** Definition of a simple point table

The different options for defining grading and point tables are summarized in Table 5.4. Under certain circumstances it might be required to change the standard wording (e.g. 'Question') as shown Tables 5.24 and 5.25. This might be the case, for example, if the language used must be changed. Table 5.5 indicates the required commands for changing the standard wording.

A simple example to change the grading table of Fig. 5.24 to German language is provided in Listing 5.24 and Fig. 5.26.

**Table 5.4**  Commands for defining grading and point tables

| Command | Comment |
|---|---|
| Grading table | |
| \gradetable[v][questions] | Vertically oriented table indexed by question number |
| \gradetable[h][questions] | Horizontally oriented table indexed by question number |
| \gradetable[v][pages] | Vertically oriented table indexed by page number |
| \gradetable[h][pages] | Horizontally oriented table indexed by page number |
| Point table | |
| \pointtable[v][questions] | Vertically oriented table indexed by question number |
| \pointtable[h][questions] | Horizontally oriented table indexed by question number |
| \pointtable[v][pages] | Vertically oriented table indexed by page number |
| \pointtable[h][pages] | Horizontally oriented table indexed by page number |

**Table 5.5**  Commands for changing the standard wording in grading and point tables

| Command | Comment |
|---|---|
| Vertical tables | |
| \vqword{text} | Replaces text for 'Question' |
| \vpgword{text} | Replaces text for 'Page' |
| \vpword{text} | Replaces text for 'Points' |
| \vsword{text} | Replaces text for 'Score' |
| \vtword{text} | Replaces text for 'Total' |
| Horizontal tables | |
| \hqword{text} | Replaces text for 'Question'. |
| \hpgword{text} | Replaces text for 'Page'. |
| \hpword{text} | Replaces text for 'Points'. |
| \hsword{text} | Replaces text for 'Score'. |
| \htword{text} | Replaces text for 'Total'. |

```
1  \begin{center}
2      \vqword{Aufgabe:}
3      \vpword{Punkte:}
4      \vtword{\textbf{Summe}}
5      \vsword{Davon erreicht:}
6    \gradetable[v][questions]
7  \end{center}
```

**Listing 5.24**  Conversion of a simple grading table to German language

**Fig. 5.26** Conversion of a
simple grading table to
German language, see
Listing 5.24

| Aufgabe: | Punkte: | Davon erreicht: |
|:---:|:---:|:---:|
| 1 | 2 | |
| 2 | 2 | |
| 3 | 4 | |
| **Summe** | 8 | |

# Chapter 6
# E-Learning: Blended Learning and Flipped Classroom Support

**Abstract**  This chapter introduces the AcroTEX eDucation Bundle which allows to generate interactive PDF documents. The focus within this context is on the exerquiz package to generate exercises, short quizzes, and quizzes. Different ways of defining questions and ways of automatically providing a feedback to the answers are introduced. The bundle also allows to display, if required, the solutions to the questions and to automatically score the provided answers.

## 6.1  The AcroTEX eDucation Bundle, Acrobat and JavaScript

The AcroTEX eDucation Bundle (AeB) can be downloaded as a single ZIP file named **acrotex.zip** and instructions on the installation, depending on the operating system used, can be found in [38]. To use the interactive PDF features, Adobe Acrobat Reader or Acrobat Pro should be used as the viewer. Important is the installation of the JavaScript file **aeb.js**. Detailed instructions for installing the aeb.js file are provided in the document **install_jsfiles.pdf**, see [35]. The correctness of the installation can be checked with the file **test_install.pdf** in the Adobe Acrobat Reader or Pro program, see [36].

## 6.2  The Exerquiz Package

The package must be included in the document preamble as shown in Listing 6.1 below:

```
1   \documentclass[...] {...}
2     \usepackage{amsmath}
3     \usepackage{exerquiz}
4   ...
5   \begin{document}
6   ...
```

**Listing 6.1**  Incorporation of the EXERQUIZ package

$$\text{Differentiate } \dfrac{\mathrm{d}}{\mathrm{d}x}\left(x^3 + 2 \times x\right) = \boxed{\phantom{xxxxxxxxxxxxxxxxxxxx}}$$

**Fig. 6.1** A simple math fill-in question, see Listing 6.2

The exerquiz package provides the following environments, see [38]:

- **exercise**: similar to the exam document class, see Chap. 5. No real interactive PDF functionality.
- **shortquiz**: used to create multiple choice questions and math/text fill-in questions with immediate response.
- **quiz**: used to create graded quizzes. In this case, several questions are bundled together. The student takes the quiz and responses are recorded by JavaScript. Upon completion of the quiz, the total score is reported to the student. The quiz environment allows to generate multiple choice questions and math/text fill-in questions. questions.
- **oQuestion**: a very simple environment for posing a *single* question (math/text fill-in).

### 6.2.1  Objective Style Questions

The exerquiz package distinguishes between two types of questions, i.e., a mathematical fill-in question that requires a mathematical answer and a text fill-in question that requires a text answer. Let us first focus on the math fill-in question. Use common notation, e.g. + (addition), - (subtraction), * (multiplication), ^ (to the power of), to define mathematical expression. Furthermore, let us illustrate the math fill-in questions within the simplest environment, i.e. the \begin{oQuestion}[name]...\end{oQuestione} environment, see Listing 6.2. The \RespBoxMath command is used define the fill-in question. The first argument of this command in Listing 6.2 provides the correct answer to the question. The second argument (4) is the random number of sample points, which is used to compare the user-provided solution proposal with the correct solution in the interval [0, 3] (fourth parameter). This numerical evaluation is based on the provided precision of 0.0001 (third parameter). The output of the compiled Listing 6.2 is shown in Fig. 6.1.

```
1  \begin{oQuestion}{diff1}
2          Differentiate $\dfrac{\text{d}}{{\text{d}x}}
3              \left(x^3+2\times x\right) =\;
4          \RespBoxMath{3*x^2+2}{4}{.0001}{[0,3]}$
5  \end{oQuestion}
```

**Listing 6.2** Environment for a simple math fill-in question

After entering any solution proposal in the answer box and pressing the 'return' key, a window with ether 'Right!' (and/or the border of the text box turns green) or

Differentiate $\dfrac{\mathrm{d}}{\mathrm{d}x}\left(x^3 + 2 \times x\right) =$ | 3*x^2+2 |

**Fig. 6.2** A simple math fill-in question with the confirmation of the correct answer

**Table 6.1** Available mathematical functions and expressions

| sin(x) | As well as other trigonometric functions, i.e. cos, tan, cot, sec, csc and the inverse functions asin, acos, atan |
|---|---|
| ln(x) | The natural logarithm |
| exp(x) | The natural exponential function, i.e. the logarithm to the base of a number |
| abs(x) | The absolute function |
| sqrt(x) | The square root function |

Integrate $\int \left(\sin(x) + 2\right)\mathrm{d}x =$ | |

**Fig. 6.3** A simple math fill-in question with a predefined function, see Listing 6.3

'Wrong!" (and/or the border of the text box turns red) is displayed, see Fig. 6.2. Not only simple mathematical expressions can be used in math fill-in questions. Some of the available functions are summarized in Table 6.1 and Listing 6.3 and Fig. 6.3 illustrates an application example.

```
1  \begin{oQuestion}{int1}
2              Integrate $\int \left(\sin(x)+2\right)\text{d}x =\;
3              \RespBoxMath{−cos(x)+2∗x}{4}{.0001}{[0,3]}$
4  \end{oQuestion}
```

**Listing 6.3** Environment for a simple math fill-in question with a predefined function

The answer of a math fill-in question does not need to be again a functions. Scalar answers are also possible, see Listing 6.4 and Fig. 6.4 for the output.

```
1  \begin{oQuestion}{add}
2              Calculate $2+3 =\;
3              \RespBoxMath{5}{1}{.0001}{[4,6]}$
4  \end{oQuestion}
```

**Listing 6.4** Environment for a simple math fill-in question with a scalar result

Calculate $2 + 3 =$ ☐

**Fig. 6.4** A simple math fill-in question with a scalar result, see Listing 6.4

It should be mentioned here that the interval [4, 6] in Listing 6.4 could be narrowed to [5, 5]. In the following, let us look a bit closer at the \RespBoxMath command. The most general structure of this command is given as follows:

\RespBoxMath[#1]{#2}(#3)[#4]{#5}{#6}{#7#8}[#9]*{#10}

The different parameters have the following meaning, see [38] for more details:

#1   Optional parameter used to modify the appearance of the answer box, see Table 6.2 for details.

#2   Correct answer to the question.

#3   An optional parameter, delimited by parentheses, that defines the independent variable. The default value is $x$. For a multivariate question, just list the variables as $(xyz)$ or $(x, y, z)$.

#4   Optional parameter which names the destination to the solution. If this parameter is set, a solution must follow in a solution environment.

#5   The number of samples points to numerically compare the entered solution with the correct one.

#6   Precision for the comparison between the entered solution and the correct one. A non-negative small value.

#7   Parameters #7 and #8 are used to define the interval from which to take the sample points. It is given in standard interval notation: $[a, b]$. For a multivariate function, the intervals can be given, for example, by $[0, 2] \times [1, 2] \times [3, 4]$.

#8   See previous item.

#9   Optional parameter which names a customized comparison function.

#10  The name of a JavaScript function that is used to process the user input. This is only detected if followed by an asterisk '*'.

Let us now have a closer look on the appearance of the answer box, i.e. the parameter #1 of the \RespBoxMath command.

An example to illustrate a modified answer box width is given in Listing 6.5 and Fig. 6.5.

```
1   \begin{oQuestion}{add}
2           Calculate $2+3 =\;
3           \RespBoxMath{5}{1}{.0001}{[4,5]}$\\
4           Calculate $2+3 =\;
5           \RespBoxMath[\rectW{1.5cm}]{5}{1}{.0001}{[4,6]}$
6   \end{oQuestion}
```

**Listing 6.5** Environment for a simple math fill-in question with modified box width

**Table 6.2** Modification of the appearance of an answer box in fill-in questions: the parameter #1 of the \RespBoxMath command

| Parameter | Explanation |
| --- | --- |
| \BC | The boundary color, a list of 0 (transparent), 1 (gray), 3 (RGB) or 4 (CMYK) numbers between 0 and 1. \BC{} results in a transparent boundary |
| \BG | Background color. The color specification is the same as in the case of \BG. \BC{} results in a transparent background |
| \S | Line style, values are S (solid), D (dashed), B (beveled), I (inset), U (underlined) |
| \rectW | The width of the answer box, e.g. \rectW{3.5cm} |
| \rectH | The height of the answer box, e.g. \rectH{2\baselineskip} or \rectH{30bp} |
| \textSize | Size in points of the text |
| \textFont | Font to be used to display the text. For example \textFont{Helv} |
| \textColor | Color of the text. There are several color spaces, including grayscale and RGB. For example, \textColor{1 0 0 rg} gives a red font |

Calculate $2 + 3 =$ [ ]

Calculate $2 + 3 =$ [ ]

**Fig. 6.5** A simple math fill-in question with modified box width, see Listing 6.5

The examples which were presented up to now allowed only to judge if a provided answer is correct or not. This was indicated by an additional window with the comment 'Right!' or 'Wrong!' and by changing the boundary color of the answer box to green or red. This means that the correct answer is not displayed to the user. To do so, one may introduce an answer key which allows to immediately display the correct answer if the used presses this key. The respective command is \CorrAnsButton and the only argument is the correct answer. Listing 6.6 and Fig. 6.6 illustrate the use of the answer button. The T<sub>E</sub>X command \kern in line 5 is used to introduce a small gap between both boxes.

```
1  \begin{oQuestion}{answer1}
2          Differentiate $\dfrac{\text{d}}{{\text{d}x}}
3                  \left(x^3+2\times x\right) =\;
4          \RespBoxMath[\rectW{2.0cm}]{3*x^2+2}{4}{.0001}{[0,3]}
5          \kern 3bp \CorrAnsButton{3*x^2+2}$
6  \end{oQuestion}
```

**Listing 6.6** Environment for a simple math fill-in question with a correct answer button

Another interesting feature is a counter for the number of wrong answers a user has tried. This can be achieved by the macro \sqTallyBox, see Listing 6.7 and Fig. 6.7 for an example.

$$\text{Differentiate } \frac{\mathrm{d}}{\mathrm{d}x}\left(x^3 + 2 \times x\right) = \boxed{\phantom{XXXXXXX}}\,\boxed{\text{Ans}}$$

**Fig. 6.6** A simple math fill-in question with a correct answer button, see Listing 6.6

$$\text{Differentiate } \frac{\mathrm{d}}{\mathrm{d}x}\left(x^3 + 2 \times x\right) = \boxed{\text{3*x+2}\phantom{XX}}\,\boxed{\text{Ans}}\,\boxed{1}$$

**Fig. 6.7** A simple math fill-in question with a counter for the number of wrong answers, see Listing 6.7

$$\text{Differentiate } \frac{\mathrm{d}}{\mathrm{d}x}\left(x^3 + 2 \times x\right) = \boxed{\phantom{XXXXXX}}\,\boxed{\text{Ans}}\,\boxed{\phantom{X}}\,\boxed{\text{Clear}}$$

**Fig. 6.8** A simple math fill-in question with a clear button, see Listing 6.8

```
1    \begin{oQuestion}{count1}
2            Differentiate $\dfrac{\text{d}}{{\text{d}x}}
3                    \left(x^3+2\times x\right) =\;
4            \RespBoxMath[\rectW{2.0cm}]{3*x^2+2}{4}{.0001}{[0,3]}
5            \kern 3bp \CorrAnsButton{3*x^2+2}\kern 3bp \sqTallyBox$
6    \end{oQuestion}
```

**Listing 6.7** Environment for a simple math fill-in question with a counter for the number of wrong answers

A final feature is a clear button which can be used to clear the answer box and the counter for the wrong answers. This can be achieved by the command \sqClearButton, see Listing 6.8 and Fig. 6.8 for an example.

```
1    \begin{oQuestion}{clear1}
2            Differentiate $\dfrac{\text{d}}{{\text{d}x}}
3                    \left(x^3+2\times x\right) =\;
4            \RespBoxMath[\rectW{2.0cm}]{3*x^2+2}{4}{.0001}{[0,3]}
5            \kern 3bp \CorrAnsButton{3*x^2+2}\kern 3bp \sqTallyBox
6            \kern 3bp \sqClearButton$
7    \end{oQuestion}
```

**Listing 6.8** Environment for a simple math fill-in question with a clear button

Let us look now at a particular math fill-in question with a vector as answer. To handle such vector responses, the dljslib package[1] can be included. The RespBoxMath command must then contain at position #10 the JavaScript function ProcVec, see Listing 6.9 and Fig. 6.9.

---

[1] In addition to vector responses, further particular mathematical answers such as equations or indefinite integrals can be handled, see [38] for details.

Instructions: Enter vectors in the solution field with angle brackets and a comma as separator, e.g., <2,3,9>.

Given are the vectors $\boldsymbol{a} = \begin{bmatrix} 1, & 4, & 2 \end{bmatrix}^{\mathrm{T}}$ and $\boldsymbol{b} = \begin{bmatrix} 1, & 2, & 5 \end{bmatrix}^{\mathrm{T}}$. Calculate the following:

$$\boldsymbol{a} + \boldsymbol{b} = \boxed{\phantom{xxxxxxxxxxxxxxxxxxxxxxxxxx}} \qquad \boxed{\text{Ans}}\boxed{\ } \boxed{\text{Clear}}$$

**Fig. 6.9** Math fill-in question to process vector responses, see Listing 6.9

```
1   ...
2   \usepackage{amsmath,amscd}
3   \usepackage{exerquiz}
4   \usepackage[vectors]{dljslib}
5
6   \renewcommand{\vec}[1]{\mbox{\boldmath$#1$}}
7
8
9   \begin{document}
10  ...
11  \noindent \textcolor{red}{Instructions:} Enter vectors in the solution field
12  with angle brackets and a comma as separator, e.g., \verb|<2,3,9>|.\\
13
14  \begin{oQuestion}{clear1}
15  \noindent Given are the vectors $\vec a =  \begin{bmatrix} 1, & 4, & 2
16  \end{bmatrix}^\text{T}$ and $\vec b =  \begin{bmatrix}1, & 2, & 5
17  \end{bmatrix}^\text{T}$. Calculate the following:\\
18
19  $\vec a + \vec b = \RespBoxMath{<2, 6, 7>}{1}{.0001}{[2,4]}*{ProcVec}$
20  \hfill\CorrAnsButton{<2, 6, 7>}\kern1bp\sqTallyBox\kern1bp\sqClearButton
21
22  \end{oQuestion}
```

**Listing 6.9** Environment for a math fill-in question to process vector responses

The second example of math fill-in questions with a vector response is related to vector function response, i.e. not a pure numerical answer, see Listing 6.10 and Fig. 6.10.

```
1   \begin{oQuestion}{vector2}
2   \noindent Given is the vector $\vec f(x) = \begin{bmatrix} 3x^2, & \cos(x),
3   & \ln(x)\end{bmatrix}^\text{T}$. Calculate the following:\\
4
5   $\vec f'(x) =
6   \RespBoxMath{<6*x, −sin(x),1/x>}(x){3}{.0001}{0}{1}*{ProcVec}$
7   \hfill\CorrAnsButton{<6*x, −sin(x),1/x>}
8   \kern1bp\sqTallyBox\kern1bp\sqClearButton
9
10  \end{oQuestion}
```

**Listing 6.10** Environment for a math fill-in question to process vector responses

Let us focus in the following on the text fill-in question. The basic command is \RespBoxTxt and the most general structure of this command is given as follows:

$$\text{\textbackslash RespBoxRespBoxTxt[\#1]\{\#2\}(\#3)[\#4]\{\#5\}\{list of alternatives\}}$$

Given is the vector $\boldsymbol{f}(x) = \begin{bmatrix} 3x^2, & \cos(x), & \ln(x) \end{bmatrix}^{\mathrm{T}}$. Calculate the following:

$\boldsymbol{f}'(x) =$ ☐☐☐☐☐☐☐☐☐☐                     [Ans☐] [Clear]

**Fig. 6.10**   Math fill-in question to process vector responses, see Listing 6.10

The different parameters have the following meaning, see [38] for more details:

**#1** Optional parameter used to modify the appearance of the answer box, see Table 6.2 for details.

**#2** This required parameter is a number that indicates the filtering method:

  **-1** The default value. No filtering al all, i.e. spaces, case, and punctuation are preserved.
  **0** Answers are converted to lower case, any white space and non-word characters are removed.
  **1** Answers are converted to lower case, any white space is removed.
  **2** Any white space in the answers is removed.

**#3** This parameter is a number that indicates the comparison method:

  **0** The default value. The author's and user's answers are compared for an exact match.
  **1** The user's response is searched in an attempt to get a substring match with the author's alternatives.

**#4** Optional parameter to name the destination of the solution. If this parameter appears, then the solution must be provided in a solution environment.

**#5** This required parameter is the number of alternative answers. All the alternative answers are listed immediately after this parameter.

Listing 6.11 and Fig. 6.11 show the example of a fill-in question with four alternative answers that are acceptable.

```
1   \begin{oQuestion}{txt1}
2               Name one of the classical numerical approximation methods\\
3               in structural mechanics:\\
4               \RespBoxTxt{0}{0}{4}{finite element method}
5               {boundary element method}{finite difference method}
6               {finite volume method}
7   \end{oQuestion}
```

**Listing 6.11**   Environment for a simple text fill-in question

An alternative way of a text fill-in question as described above is the text question with partial credit: each time one of the keywords are found in the input string, a predefined credit for each word is given. This allows to enter a longer text which is screened for the 'correct' words. The corresponding command is \RespBoxTxtPC and the most general structure of this command is given as follows:

\RespBoxRespBoxTxtPC[#1]{#2}[#3]{#4}[num1]{word1}...[num_n]{word_n}

Name one of the classical numerical approximation methods
in structural mechanics:

**Fig. 6.11** A simple text fill-in question, see Listing 6.11

Which French foreign minister proposed in 1950 the European
Coal and Steel Community (ECSC)?

**Fig. 6.12** A simple text fill-in question with partial credit, see Listing 6.12

The different parameters have the following meaning, see [38] for more details:

#1    Optional parameter used to modify the appearance of the answer box, see
Table 6.2 for details.

#2    This required parameter is a number that indicates the filtering method:

-1    The default value. No filtering al all, i.e. spaces, case, and punctuation are
preserved.

0    Answers are converted to lower case, any white space and non-word
characters are removed.

1    Answers are converted to lower case, any white space is removed.

2    Any white space in the answers is removed.

3    Same as -1, but a case insensitive search is performed (recommended).

#3    Optional parameter to name the destination of the solution. If this parameter
appears, then the solution must be provided in a solution environment.

#4    This required parameter is the number of alternative answers that are accept-
able. The alternative answers are listed immediately after this parameter and
are of the form [num]word. The [num] is the amount of credit the user gets if
his answer contains the specified 'word'.

A question is judged correct (and the border of the answer box turns green) if at least
one of the words is found. Listing 6.12 and Fig. 6.12 show a simple example of a
fill-in question with partial credit.

```
1  \begin{oQuestion}{txtPC1}
2      Which French foreign minister proposed in 1950 the European\\
3          Coal and Steel Community (ECSC)?\\
4      \RespBoxTxtPC{3}{2}
5      [1]{Robert}
6      [2.5]{Schuman}
7  \end{oQuestion}
```

**Listing 6.12** Environment for a simple text fill-in question with partial credit

In the example of Listing 6.12, one point would be awarded for the first name 'Robert'
and 2.5 points for the family name 'Schuman'. If someone would have misspelled

the family name as 'Schumann', the same credit of 2.5 points would have been awarded since the string 'Schuman' would have been found. To avoid this mistake, it is possible to clearly indicate the word boundaries with the command[2] \\b, see Listing 6.13 for details.

```
1   \begin{oQuestion}{txtPC1}
2           Which French foreign minister proposed in 1950 the European\\
3                   Coal and Steel Community (ECSC)?\\
4           \RespBoxTxtPC{3}{2}
5           [1]{\\bRobert{\\b}
6           [2.5]{{\\bSchuman{\\b}
7   \end{oQuestion}
```

**Listing 6.13** Environment for a simple text fill-in question with partial credit and indication of word boundaries

Let us mention in the following a way to state several alternative versions of a solution or to include abbreviations. This can be done by grouping each alternative in parentheses and the | character is the symbol for an alternation, see Listing 6.14 for details.

```
1   \begin{oQuestion}{txtPC1}
2           Which French foreign minister proposed in 1950 the European\\
3                   Coal and Steel Community (ECSC)?\\
4           \RespBoxTxtPC{3}{2}
5           [1]{(\\bRobert\\b|\\bRob.{0,1}\\b|\\bR.{0,1}\\b)}
6           [2.5]{Schuman}
7   \end{oQuestion}
```

**Listing 6.14** Environment for a simple text fill-in question with partial credit and alternative solution words

In the example of Listing 6.14, 'Robert', 'Rob.', 'Rob', 'R.', and 'R' would have been counted as a correct answer for the first name. The {0,1} following the period (.) means match 0 or 1 period.

Spaces in a solution word can be problematic. If, for example, the solution word should be 'Richard von Mises', it is better to use the \\s+ command to indicate the space. Thus, the solution should be mentioned in the \RespBoxTxtPC environment as Richard\\s+von\\s+Mises.

In the previous elaborations it was explained that a user will get a message indicating if his answer is right or wrong. The summation of the partial credits and the total number of obtained points was omitted. To obtain this information, it is required to move from the oQuestion environment to the quiz environment,[3] which reveals much more options. Including the \PointsField command allows to display the sum of the obtained points, see Listing[4] 6.15 and Fig. 6.13.

---

[2] An alias for the \\b...\\b command sequence is \word{...}.

[3] See Sect. 6.2.2 for more details on the quiz environment.

[4] It turned out that the source code must contain the commands \useBeginQuizButton and \useEndQuizButton before the \begin{document} declaration. Otherwise, the summation of the partial credits does not work properly.

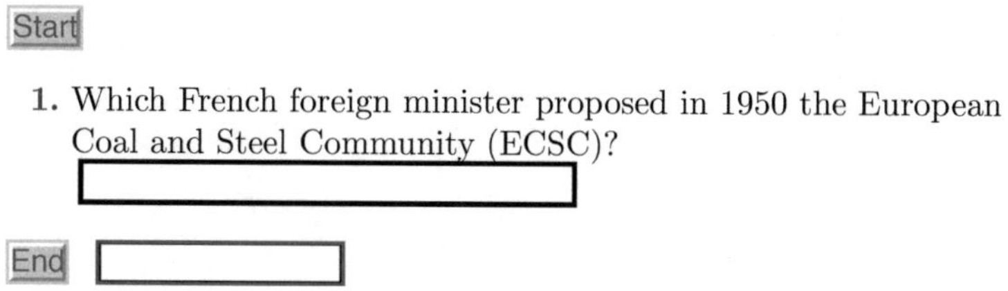

Start

    **1.** Which French foreign minister proposed in 1950 the European
       Coal and Steel Community (ECSC)?

End

**Fig. 6.13** A text fill-in question within a quiz environment, see Listing 6.15

```
1  \begin{quiz}{qu1}
2      \begin{questions}
3          \item\PTs{3} Which French foreign minister proposed in 1950 the
4                       European\\Coal and Steel Community (ECSC)?\\
5          \RespBoxTxtPC{3}{2}[1]{Robert}[2]{Schuman}
6      \end{questions}
7  \end{quiz}\quad\PointsField[\rectW{2.5cm}]\currQuiz
```

**Listing 6.15** Environment for a text fill-in question within a quiz environment

It should be noted here that a user must first press the 'Start' button (see Fig. 6.13) before filling in the solution and the end must be indicated by pressing the 'End' button. After this, the total score will be displayed.

The final command for text fill-in questions is \RespBoxEssay. This allows to provide an extended response to a question. However, the answer is not automatically evaluated and must be reviewed by the instructor. The most general structure of this command is given as follows:

$$\text{\textbackslash RespBoxEssay[options]dest\{wd\}\{ht\}}$$

The different parameters have the following meaning, see [38] for more details:

| | |
|---|---|
| options | Optional parameter used to pass key-value pairs to the form field. |
| dest | Optional parameter: (1) a named destination to the solution or (2) an asterisk, where the named destination is automatically generated by exerquiz. |
| wd | The width of the multi-line field. |
| ht | The height of the multi-line field. |

An example of this environment is provided in Listing 6.16 and Fig. 6.14.

```
1  \begin{oQuestion}{essay1}
2      State the most significant milestones to establish the European
3          Union (EU):
4          \RespBoxEssay[]{\textwidth}{3cm}
5  \end{oQuestion}
```

**Listing 6.16** Environment for a text fill-in question for extended response (\RespBoxEssay)

State the most significant milestones to establish the European Union (EU):

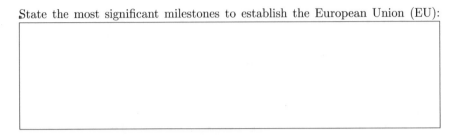

**Fig. 6.14**  A text fill-in question for an extended response (\RespBoxEssay), see Listing 6.16

### 6.2.2   The quiz Environment

The quiz environment is used to create a collection of graded questions, i.e. multiple choice questions and math/text fill-in questions. The responses are automatically recorded and the total score can be provided upon completion of the quiz. The general structure of the quiz environment is shown in Listing 6.17, see [38] for more details.

```
1   \begin{quiz}{name}
2       Some initial text.
3       \begin{questions}
4           \item State first question ...
5       \begin{answers}{4} % <- num_cols = 4
6           \Ans0 ... &\Ans1 ... &\Ans0 ... & \Ans0 ...
7       \end{answers}
8       ...
9           \item State n-th question ...
10      \begin{answers}{4} % <- 4 column format
11          \Ans0 ... &\Ans1 ... &\Ans0 ... &\Ans0 ...
12      \end{answers}
13      \end{questions}
14  \end{quiz}
```

**Listing 6.17**  The general structure of the quiz environment

Place the command \ScoreField{name} immediately after the quiz environment to obtain the total score, i.e., the number of correct questions.[5] Some comments on Listing 6.17: The answers environment requires the number of columns (here: 4) to typeset the answers in a tabular environment. If a '1' is used, the answers are given in a list environment. It should be highlighted here that only one correct answer can be provided within a single answer environment. The command \Ans has two possile arguments. A '0' to indicate that the answer is wrong and a '1' to indicate the correct answer. A quiz has normally a Begin Quiz and a End Quiz button which must be pressed by the user at the beginning and to complete the quiz, i.e. to sum up the partial points. The standard appearance can be easily modified as illustrated in Listing 6.18

---

[5] The command \ScoreField{name} should not be confused with the command \PointsField. The first one gives the sum of correct answers, the second one the sum of achieved points.

```
1  ...
2  \useBeginQuizButton[\BC{}\textColor{1 0 0}\CA{Start}]
3  \useEndQuizButton[\BC{}\textColor{1 0 0}\CA{End}]
4  ...
5  \begin{document}
```

**Listing 6.18** Mofification of the appearance of the Begin Quiz and End Quiz buttons

A simple example of a quiz in 'checkbox' format is shown in Listing 6.19 and Fig. 6.15. The 'checkbox' format is obtained using the **quiz\*** command.

```
1   \begin{quiz*}{quiz01}
2       Respond to each of the following questions.
3       \begin{questions}
4       \item What is the Young's modulus of steel?
5       \begin{answers}{4}
6       \Ans0 70000 MPa &\Ans1 210000 MPa  &\Ans0 110000 MPa
7       \end{answers}
8       \item What is the Poisson's ratio of steel?
9       \begin{answers}{4}
10      \Ans0 0.34 &\Ans0 0.5 &\Ans1 0.3
11      \end{answers}
12      \end{questions}
13  \end{quiz*}\par
14  \ScoreField{quiz01}
```

**Listing 6.19** Environment for a simple quiz in 'checkbox' format

The classical **answers** environment is limited to a single correct answer. In case that more than one answer is correct, the **manswers** environment (i.e., multiple answers) must be used, see Listing 6.20 and Fig. 6.16 for an example. The **manswers** environment requires, as in the case of the **answers** environment, the number of columns (here: 4) to typeset the answers in a tabular environment. If a '1' is used, the answers are given in a single list environment.

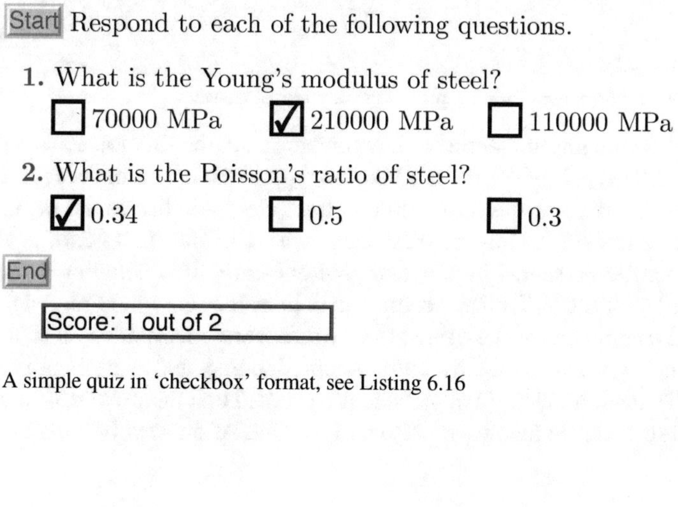

**Fig. 6.15** A simple quiz in 'checkbox' format, see Listing 6.16

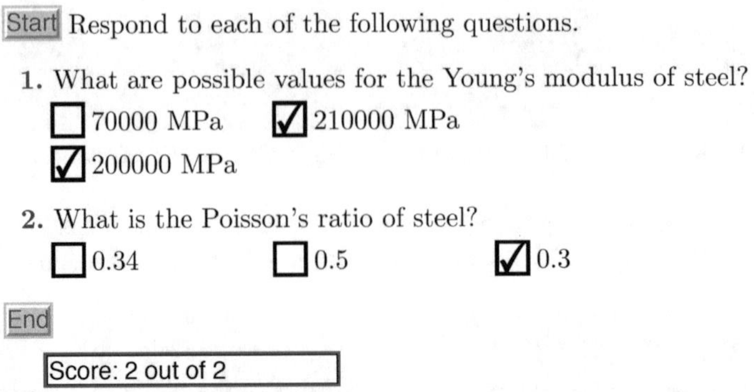

Start Respond to each of the following questions.

1. What are possible values for the Young's modulus of steel?

☐ 70000 MPa        ☑ 210000 MPa

☑ 200000 MPa

2. What is the Poisson's ratio of steel?

☐ 0.34            ☐ 0.5            ☑ 0.3

End

Score: 2 out of 2

**Fig. 6.16**  A quiz with a multiple answers environment, see Listing 6.20

```
1   \begin{quiz*}{quiz02}
2   Respond to each of the following questions.
3       \begin{questions}
4           \item What are possible values for the Young's modulus of steel?
5           \begin{manswers}{4}
6               \rowsep{3pt}
7               \bChoices[2]
8               \Ans{0}  70000 MPa \eAns
9               \Ans{1} 210000 MPa \eAns
10              \Ans{1} 200000 MPa \eAns
11              \eChoices
12          \end{manswers}
13          \item What is the Poisson's ratio of steel?
14          \begin{answers}{4}
15              \bChoices[]
16              \Ans{0}  0.34  \eAns
17              \Ans{0}  0.5   \eAns
18              \Ans{1}  0.3   \eAns
19              \eChoices
20          \end{answers}
21      \end{questions}
22  \end{quiz*}\par
23  \ScoreField{quiz02}
```

**Listing 6.20**  Environment for a simple quiz with multiple answers

Listing 6.20 contains another new 'environment', i.e. the \bChoices ... \eChoices commands. This pair of commands helps to typeset the list of choices for a multiple choice question. The \Ans ... \eAns construct clearly indicates the beginning and the end of a choice. Listing 6.20 indicates with c of the manswers environment that choices are presented in a tabular environment with 4 columns. However, the argument '2' of the \bChoices command defines that only the first 2 columns are used. If the argument of the answers or manswers environment is changed to 1, the optional argument of \bChoices is simply ignored, and everything is typeset in a single list environment. Finally, it should be noted that the \rowsep{...} command was used to increase the space between two rows of choices (the default value is 0 pt).

Start Respond to each of the following questions.

1. What are possible values for the Young's modulus of steel?
   ☒ 70000 MPa        ☑ 210000 MPa
   ◉ 200000 MPa

2. What is the Poisson's ratio of steel?
   ☐ 0.34              ☐ 0.5              ☑ 0.3

End

| Score: 1 out of 2 | | Correct |

**Fig. 6.17** A quiz with a correction button, see Listing 6.21

In all the previous examples, the user could get feedback based on the \ScoreField{...} command, i.e., the info on the total number of correct answers. However, feedback on which particular answer was correct or wrong was not provided. This feedback can be provided by using the macro \eqButton{name}, see Listing 6.21 and Fig. 6.17 for an example. A ✔ indicates a correct response, a ✗ indicates an incorrect response, and the correct answer is indicated with a ●.

```
1    \begin{quiz*}{quiz03}
2    Respond to each of the following questions.
3        \begin{questions}
4            \item What are possible values for the Young's modulus of steel?
5            \begin{manswers}{4}
6                \rowsep{3pt}
7                \bChoices[2]
8                \Ans{0}  70000 MPa \eAns
9                \Ans{1} 210000 MPa \eAns
10               \Ans{1} 200000 MPa \eAns
11               \eChoices
12           \end{manswers}
13           \item What is the Poisson's ratio of steel?
14           \begin{answers}{4}
15               \bChoices[]
16               \Ans{0} 0.34  \eAns
17               \Ans{0} 0.5   \eAns
18               \Ans{1} 0.3   \eAns
19               \eChoices
20           \end{answers}
21       \end{questions}
22   \end{quiz*}\par
23   \ScoreField{quiz03}\quad \eqButton{quiz03}
```

**Listing 6.21** Environment for a quiz with a correction button

Let us focus in the following on the question on how to assign points to questions. The following two macros can be used to define points, see [38] for details:

- \item\PTs{5}: This syntax assigns 5 points to the question which follows the \item command.
- \PTsHook{#1}: This macro is used to display the points assigned and is called by \PTs. The value assigned to the current question by \PTs is contained within the macro \eqPTs. Example: \PTsHook{($\eqPTs^{\text{pts}}$) . Omit this macro if the points should not be displayed for each single question.

To display and evaluate the obtained points,[6] the following commands can be used, see [38] for details:

- \PointsField[#1]#2: The number of points earned and the total points for the quiz. The parameter #2 is the name of the quiz.
- \PercentField[#1]#2: The percentage of points. The parameter #2 is the name of the quiz.
- \GradeField[#1]#2: The letter grade of the performance. The parameter #2 is the name of the quiz. The values placed in this field are determined by the macro \eqGradeScale.

  - \eqGradeScale: This macro defines the grade scale of a quiz. The default setting is: \newcommand\eqGradeScale{"A",[90, 100],"B",[80,90],"C",[70,80],"D",[60,70],"F",[0,60]}.

Listing 6.22 and Fig. 6.18 illustrate the incorporation of points.

```
1   \PTsHook{($\eqPTs^{\text{pts}}$)}
2   \begin{quiz*}{quiz04}
3   Respond to each of the following questions.
4       \begin{questions}
5       \item\PTs{5} What are possible values for the Young's modulus of steel?
6       \begin{manswers}{4}
7           \rowsep{3pt}
8           \bChoices[2]
9           \Ans{0}   70000 MPa  \eAns
10          \Ans{1} 210000 MPa  \eAns
11          \Ans{1} 200000 MPa  \eAns
12          \eChoices
13      \end{manswers}
14      \item\PTs{3} What is the Poisson's ratio of steel?
15      \begin{answers}{4}
16          \bChoices[]
17          \Ans{0}  0.34  \eAns
18          \Ans{0}  0.5   \eAns
19          \Ans{1}  0.3   \eAns
20          \eChoices
21      \end{answers}
22      \end{questions}
23  \end{quiz*}\quad
24  \ScoreField{quiz04}\quad\eqButton{quiz04}
25
26  \medskip\noindent
27  Points: \PointsField{quiz04}\ Percent: \PercentField{quiz04}
```

**Listing 6.22**  Environment for a quiz under consideration and evaluation of points

---

[6] The command \negPointsAllowed must be used in the preamble if negative points for wrong answers should be considered. Otherwise, any negative number will be converted into a point score of zero (default setting).

Start Respond to each of the following questions.

1. ($5^{\text{pts}}$) What are possible values for the Young's modulus of steel?
☐ 70000 MPa   ☑ 210000 MPa
☑ 200000 MPa

2. ($3^{\text{pts}}$) What is the Poisson's ratio of steel?
☐ 0.34   ☑ 0.5   ☐ 0.3

End | Score: 1 out of 2 | | Correct

Points: | Score: 5 out of 8 | Percent: | 62.5% |

**Fig. 6.18** A quiz under consideration and evaluation of points, see Listing 6.22

Let us focus now on the possibilities to provide solutions to quizzes. This can be done by including a \begin{solution}...\end{solution} environment after each \end{answers} or \end{manswers} command. This will result in the content of these environments to be shown at the end of the quiz, with each single solution on a single page. It is important here to use the optional parameter of the **manswers** and **answers** environments (here: Young and Poisson). If this parameter is not used, the solution will be shown in the **questions** environment. To avoid that the solutions are accessible before the quiz is completed, the command \NoPeeking can be used in the preamble or prior to a quiz. Listing 6.23 shows a corresponding example and Figs. 6.19 and 6.20 show the solutions, which are presented on the final pages.

## Solutions to Quizzes

**Solution to Quiz:** The Young's modulus of steel is in the following range: 190000 ... 210000 MPa.

■

**Fig. 6.19** A quiz with solutions for each single question (page before last), see Listing 6.23

**Solution to Quiz:** Poisson's ratio of steel is approximately 0.3.

■

**Fig. 6.20** A quiz with solutions for each single question (last page), see Listing 6.23

```
1   \NoPeeking
2   \PTsHook{($\eqPTs^{\text{pts}}$)}
3   \begin{quiz*}{quiz05}
4    Respond to each of the following questions.
5       \begin{questions}
6       \item\PTs{5} What are possible values for the Young's modulus of steel?
7       \begin{manswers}[Young]{4}
8          \rowsep{3pt}
9          \bChoices[2]
10         \Ans{0}  70000 MPa \eAns
11         \Ans{1} 210000 MPa \eAns
12         \Ans{1} 200000 MPa \eAns
13         \eChoices
14      \end{manswers}
15      \begin{solution}
16        The Young's modulus of steel is in the following range:
17              190000 ... 210000 MPa.
18      \end{solution}
19      \item\PTs{3} What is the Poisson's ratio of steel?
20      \begin{answers}[Poisson]{4}
21         \bChoices[]
22         \Ans{0} 0.34 \eAns
23         \Ans{0} 0.5  \eAns
24         \Ans{1} 0.3  \eAns
25         \eChoices
26      \end{answers}
27      \begin{solution}
28        Poisson's ratio of steel is approximately 0.3.
29      \end{solution}
30      \end{questions}
31   \end{quiz*}\quad
32   \ScoreField{quiz05}\quad\eqButton{quiz05}
33
34   \medskip\noindent
35   Points: \PointsField{quiz05}\ Percent: \PercentField{quiz05}
```

**Listing 6.23**  Environment for a quiz with solutions for each single question

To collect all the solutions on a single page, a single \begin{solution}...\end{solution} environment can be introduced before the final \end{quiz*} command. This environment should now contain all the answers to each of the single questions, see Listing 6.24 and the corresponding output in Fig. 6.21.

### Solutions to Quizzes

**Solution to Quiz:** The Young's modulus of steel is in the following range: 190000 ... 210000 MPa. Poisson's ratio of steel is approximately 0.3. These values are valid at moderate temperatures.

■

**Fig. 6.21**  A quiz with all solutions on the final page, see Listing 6.24

```
1   \NoPeeking
2   \PTsHook{($\eqPTs^{\text{pts}}$)}
3   \begin{quiz*}{quiz06}
4   Respond to each of the following questions.
5       \begin{questions}
6       \item\PTs{5} What are possible values for the Young's modulus of steel?
7       \begin{manswers}[Young]{4}
8           \rowsep{3pt}
9           \bChoices[2]
10          \Ans{0}  70000 MPa \eAns
11          \Ans{1} 210000 MPa \eAns
12          \Ans{1} 200000 MPa \eAns
13          \eChoices
14      \end{manswers}
15      \item\PTs{3} What is the Poisson's ratio of steel?
16      \begin{answers}[Poisson]{4}
17          \bChoices[]
18          \Ans{0} 0.34 \eAns
19          \Ans{0} 0.5  \eAns
20          \Ans{1} 0.3  \eAns
21          \eChoices
22      \end{answers}
23      \end{questions}
24      \begin{solution}
25          The Young's modulus of steel is in the following range:
26          190000 ... 210000 MPa.
27          Poisson's ratio of steel is approximately 0.3.
28          These values are valid at moderate temperatures.
29      \end{solution}
30  \end{quiz*}\quad
31  \ScoreField{quiz06}\quad\eqButton{quiz06}
32
33  \medskip\noindent
34  Points: \PointsField{quiz06}\ Percent: \PercentField{quiz06}
```

**Listing 6.24**  Environment for a quiz with all solutions on the final page

If the solutions should not be shown at all in a particular version of the document, one may use the argument **noquizsolutions** of the **exerquiz** package, see Listing 6.25. Another useful option of the package is **solutionsafter**, which would print the solutions immediately after the question and not at the end of the document.

```
1   \documentclass{article}
2   ...
3   \usepackage[noquizsolutions]{exerquiz}
```

**Listing 6.25**  Hiding the solutions of quizzes

Let us look at the end of this section on an example, which contains different types of questions. Listing 6.26 and Fig. 6.22 contain a text fill-in, a multiple choice, and a math fill-in question.

```
1   \NoPeeking
2   \PTsHook{($\eqPTs^{\text{pts}}$)}
3   \begin{quiz*}{quiz07}
4   Respond to each of the following questions.
5       \begin{questions}
6       \item\PTs{3} Which French foreign minister proposed in 1950 the European\\
7           Coal and Steel Community (ECSC)?\\
8       \RespBoxTxtPC{3}|ECSC|{2}
9           [1]{Robert}
10          [2]{Schuman}
```

```
11   \CorrAnsButton{Robert Schuman}
12   \begin{solution}
13     Robert Schuman was the French foreign minister who proposed in 1950
14     the European Coal and Steel Community (ECSC).
15   \end{solution}
16   \item\PTs{3} What is the Poisson's ratio of steel?
17   \begin{answers}[Poisson]{4}
18       \bChoices[]
19       \Ans{0} 0.34  \eAns
20       \Ans{0} 0.5   \eAns
21       \Ans{1} 0.3   \eAns
22       \eChoices
23   \end{answers}
24   \begin{solution}
25     Poisson's ratio of steel is approximately 0.3.
26     This value is valid at moderate temperatures.
27   \end{solution}
28   \item\PTs{2}   Calculate $2+3 =\;
29   \RespBoxMath[\rectW{1.5cm}]{5}[sum]{1}{.0001}{[4,5]}$\\
30   \CorrAnsButton{5}
31   \begin{solution}
32     Correct calculation: $2+3=5$.
33   \end{solution}
34   \end{questions}
35 \end{quiz*}\quad
36 \ScoreField{quiz07}\quad\eqButton{quiz07}
37
38 \noindent
39 Answers: \AnswerField{quiz07}
40
41 \medskip\noindent
42 Points: \PointsField{quiz07}\ Percent: \PercentField{quiz07}
```

**Listing 6.26**  Environment for a quiz with different types of questions

Start Respond to each of the following questions.

1. ($3^{pts}$) Which French foreign minister proposed in 1950 the European Coal and Steel Community (ECSC)?

2. ($3^{pts}$) What is the Poisson's ratio of steel?
   ☐ 0.34        ☐ 0.5        ☐ 0.3

3. ($2^{pts}$) Calculate $2 + 3 =$

End  Score:                    Correct
Answers:
Points:                    Percent:

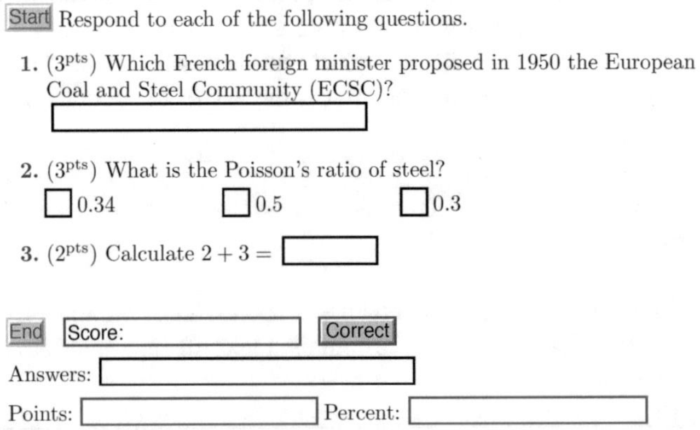

**Fig. 6.22**  A quiz with different types of questions, see Listing 6.26

Start Respond to each of the following questions.

1. ($3^{pts}$) Which French foreign minister proposed in 1950 the European
Coal and Steel Community (ECSC)?

| Robert | Ans |

2. ($3^{pts}$) What is the Poisson's ratio of steel?

☐ 0.34        ☒ 0.5           ◉ 0.3

3. ($2^{pts}$) Calculate $2 + 3 =$ [5    ]

Ans

End [Score: 2 out of 3    ]  [Correct]

Answers: [Robert Schuman        ]

Points: [Score: 3 out of 8    ]  Percent: [37.5%        ]

**Fig. 6.23** A quiz with different types of questions and solution highlighting, see Listing 6.26

Let us now have a closer look at the code provided in Listing 6.26. It is important
to highlight that each answer environment (i.e., **RespBoxTxtPC, answers**, and
**RespBoxMath**) has now its optional parameter (here: ECSC, Poisson, sum). This
is important to ensure that the solution is not shown in the problem statement. In
addition, we include the command **\AnswerField** (see line 39). This creates a text
field and, in conjunction with the command **\CorrAnsButton{...}** (see lines 11 and
30), allows to display the correct answers once the quiz is completed, see Fig. 6.23.
After finishing the quiz (**End** button) and pressing the **Correct** button, an **Ans** button
appears after the text fields of question 1 and 3. Pressing the corresponding **Ans** button
provides the solution in the **Answers** text field.

### 6.2.3 The shortquiz Environment

The **shortquiz** environment is used to create multiple choice questions and math-
/text fill-in questions with *immediate* response (and not after the completion of all
questions as in the case of the **quiz** environment). Listing 6.27 and Fig. 6.24 show
a simple shortquiz in 'checkbox' format, which assembles many of the environments and commands already explained in the previous sections. Selecting one of
the answer options gives an immediate response as illustrated in Fig. 6.24.

Quiz Respond to each of the following questions.

1. What are possible values for the Young's modulus of steel?

☐ 70000 MPa      ☑ 210000 MPa

☐ 200000 MPa

2. What is the Poisson's ratio of steel?

☐ 0.34                ☐ 0.5                ☐ 0.3

**Fig. 6.24**  A simple shortquiz in 'checkbox' format, see Listing 6.27

```
1    \begin{shortquiz*}[shortquiz01]
2    Respond to each of the following questions.
3        \begin{questions}
4        \item What are possible values for the Young's modulus of steel?
5        \begin{manswers}[4]
6            \rowsep{3pt}
7            \bChoices[2]
8            \Ans{0}  70000 MPa \eAns
9            \Ans{1} 210000 MPa \eAns
10           \Ans{1} 200000 MPa \eAns
11           \eChoices
12       \end{manswers}
13       \item What is the Poisson's ratio of steel?
14       \begin{answers}[Poisson]{4}
15           \bChoices[]
16           \Ans{0} 0.34 \eAns
17           \Ans{0} 0.5  \eAns
18           \Ans{1} 0.3  \eAns
19           \eChoices
20       \end{answers}
21       \begin{solution}
22           Poisson's ratio of steel is approximately 0.3.
23           This value is valid at moderate temperatures.
24       \end{solution}
25       \end{questions}
26   \end{shortquiz*}
```

**Listing 6.27**  Environment for a simple shortquiz in 'checkbox' format

To display the solution immediately after the questions, the command \SolutionsAfter can be used, see Listing 6.28 and Fig. 6.25. The command \SolutionsAtEnd immediately after the end of the quiz (\end{shortquiz*}) causes that the solutions of all following quizzes are again shown at the end of the document.

*Quiz* Respond to each of the following questions.

1. What are possible values for the Young's modulus of steel?
   ☐ 70000 MPa    ☐ 210000 MPa
   ☐ 200000 MPa

2. What is the Poisson's ratio of steel?
   ☐ 0.34          ☐ 0.5          ☐ 0.3

   *Solution*: Poisson's ratio of steel is approximately 0.3. This value is valid at moderate temperatures. ■

**Fig. 6.25** A simple shortquiz in 'checkbox' format with the solution immediately after the questions, see Listing 6.28

```
\SolutionsAfter
\begin{shortquiz*}[shortquiz02]
Respond to each of the following questions.
    \begin{questions}
    \item What are possible values for the Young's modulus of steel?
    \begin{manswers}{4}
        \rowsep{3pt}
        \bChoices[2]
        \Ans{0} 70000 MPa  \eAns
        \Ans{1} 210000 MPa \eAns
        \Ans{1} 200000 MPa \eAns
        \eChoices
    \end{manswers}
    \item What is the Poisson's ratio of steel?
    \begin{answers}[Poisson]{4}
        \bChoices[]
        \Ans{0} 0.34 \eAns
        \Ans{0} 0.5  \eAns
        \Ans{1} 0.3  \eAns
        \eChoices
    \end{answers}
    \begin{solution}
        Poisson's ratio of steel is approximately 0.3.
        This value is valid at moderate temperatures.
    \end{solution}
    \end{questions}
\end{shortquiz*}
\SolutionsAtEnd
```

**Listing 6.28** Environment for a simple shortquiz in 'checkbox' format with the solution immediately after the questions

Another useful command in the same context allows placing the solutions at any position in the document after the last quiz.[7] Use the command \includequizsolutions to indicate the desired location for the solutions.

Let us look at the end of this section on the modification of some of the form elements. Listing 6.29 contains the command \renewcommand\sqlabel{...} to redefine the title for the shortquiz environment, the command \titleQuiz{...} to assign a title

---

[7] This command works with the shortquiz and quiz environment.

to the shortquiz, and the command \fancyQuizHeaders to change the labeling of
the quiz in the solutions section. Furthermore, a label has been defined which allows
cross-referencing, see Figs. 6.26 and 6.27.

```
 1  \renewcommand\sqlabel{\textcolor{blue}{Questions:}}
 2  \titleQuiz{Material Parameters}
 3  \fancyQuizHeaders
 4  \begin{shortquiz*}[shortquiz03]\label{q:03}
 5  Respond to all questions.
 6      \begin{questions}
 7      \item What are possible values for the Young's modulus of steel?
 8      \begin{manswers}{4}
 9          \rowsep{3pt}
10          \bChoices[2]
11          \Ans{0} 70000 MPa  \eAns
12          \Ans{1} 210000 MPa \eAns
13          \Ans{1} 200000 MPa \eAns
14          \eChoices
15      \end{manswers}
16      \item What is the Poisson's ratio of steel?
17      \begin{answers}[Poisson]{4}
18          \bChoices[]
19          \Ans{0} 0.34 \eAns
20          \Ans{0} 0.5  \eAns
21          \Ans{1} 0.3  \eAns
22          \eChoices
23      \end{answers}
24      \begin{solution}
25          Poisson's ratio of steel is approximately 0.3.
26          This value is valid at moderate temperatures.
27      \end{solution}
28      \end{questions}
29  \end{shortquiz*}
30
31  \noindent As we questioned in Quiz~\ref{q:03}, ..
```

**Listing 6.29** Environment for a simple shortquiz in 'checkbox' format with modified form elements

Questions: **Material Parameters** Respond to all questions.

1. What are possible values for the Young's modulus of steel?
   ☐ 70000 MPa    ☐ 210000 MPa
   ☐ 200000 MPa

2. What is the Poisson's ratio of steel?
   ☐ 0.34         ☐ 0.5          ☐ 0.3

As we questioned in Quiz ☐1☐, ..

**Fig. 6.26** A simple shortquiz in 'checkbox' format with modified form elements, see Listing 6.29

## Solutions to Quizzes

**Material Parameters: Question 2.** Poisson's ratio of steel is approximately 0.3. This value is valid at moderate temperatures.

■

**Fig. 6.27** A simple shortquiz in 'checkbox' format with modified form elements (solution section), see Listing 6.29

### 6.2.4  The exercise Environment

The **exercise** environment allows to create questions (exercises) with solutions. However, the interactive PDF component—as in the case of quizzes or short quizzes—is not available. Let us try to reproduce the example shown in Fig. 6.22, i.e., a quiz with different types of questions.

```
...
\usepackage[nosolutions]{exerquiz}
%\usepackage[solutionsafter]{exerquiz}
...
\begin{document}
...
\PTsHook{($\eqPTs^{\text{pts}}$)}
\begin{exercise}
 Respond to each of the following questions.
    \begin{questions}
    \item\PTs{3} Which French foreign minister proposed in 1950 the European\\
    Coal and Steel Community (ECSC)?\\
    \begin{solution}[1cm]
        Robert Schuman was the French foreign minister who proposed in 1950
        the European Coal and Steel Community (ECSC).
    \end{solution}
    \item\PTs{3} What is the Poisson's ratio of steel?
    \begin{tabbing}
    \hspace*{2cm}\=\hspace*{2cm}\= \kill
    $\square$ 0.34 \> $\square$ 0.5 \> $\square$ 0.3 \\
    \end{tabbing}
    \begin{solution}[]
        Poisson's ratio of steel is approximately 0.3.
        This value is valid at moderate temperatures.
    \end{solution}
    \item\PTs{2}   Calculate $2+3 =\;$
    \begin{solution}[]
        Correct calculation: $2+3=5$.
    \end{solution}
    \end{questions}
\end{exercise}
```

**Listing 6.30**  Environment for an exercise with different form elements

Listing 6.30 shows that we still can use the **exercise** environment, including the **\item** command. However, the **RespBoxTxtPC** and **RespBoxMath** commands, as well as the **answers** (or **manswers**) environments are no longer applicable. Solutions are enclosed in the same environment as in the previous section. As we can see in

EXERCISE 1. Respond to each of the following questions.

1. ($3^{pts}$) Which French foreign minister proposed in 1950 the European Coal and Steel Community (ECSC)?

2. ($3^{pts}$) What is the Poisson's ratio of steel?

☐ 0.34          ☐ 0.5          ☐ 0.3

3. ($2^{pts}$) Calculate $2 + 3 =$

**Fig. 6.28**  An exercise with different form elements and hidden solutions, see Listing 6.30

EXERCISE 1. Respond to each of the following questions.

1. ($3^{pts}$) Which French foreign minister proposed in 1950 the European Coal and Steel Community (ECSC)?

   *Solution*: Robert Schuman was the French foreign minister who proposed in 1950 the European Coal and Steel Community (ECSC).                     ☐

2. ($3^{pts}$) What is the Poisson's ratio of steel?

   ☐ 0.34          ☐ 0.5          ☐ 0.3

   *Solution*: Poisson's ratio of steel is approximately 0.3. This value is valid at moderate temperatures.                     ☐

3. ($2^{pts}$) Calculate $2 + 3 =$
   *Solution*: Correct calculation: $2 + 3 = 5$.                     ☐

**Fig. 6.29**  An exercise with different form elements and shown solutions, see Listing 6.30

line 13, there is the optional parameter[8] [1cm] used which inserts a vertical space for the user to write in an answer, see Fig. 6.28. Using the option **solutionsafter** of the **exerquiz** package, prints the solutions immediately after the corresponding questions, see Fig. 6.29.

---

[8] This space is provided if the option **nosolutions** is used for the **exerquiz** package.

EXERCISE 1. Respond to each of the following questions.

(a) (3^pts) Which French foreign minister proposed in 1950 the European Coal and Steel Community (ECSC)?

(b) (3^pts) What is the Poisson's ratio of steel?

    □ 0.34       □ 0.5       □ 0.3

(c) (2^pts) Calculate $2 + 3 =$

**Fig. 6.30** An exercise with a parts environment for multiple parts questions, see Listing 6.31

Let us now focus on some particular features of the **exercise** environment. The **exercise\*** environment in conjunction with the **parts** environment allow to formulate questions with multiple parts, see Listing 6.31 and Fig. 6.30.

```
1    ...
2    \usepackage[nosolutions]{exerquiz}
3    %\usepackage[solutionsafter]{exerquiz}
4    ...
5    \begin{document}
6    ...
7    \PTsHook{($\eqPTs^{\text{pts}}$)}
8    \begin{exercise*}
9    Respond to each of the following questions.
10       \begin{parts}
11       \item\PTs{3} Which French foreign minister proposed in 1950 the European\\
12       Coal and Steel Community (ECSC)?\\
13       \begin{solution}[1cm]
14           Robert Schuman was the French foreign minister who proposed in 1950
15           the European Coal and Steel Community (ECSC).
16       \end{solution}
17       \item\PTs{3} What is the Poisson's ratio of steel?
18       \begin{tabbing}
19       \hspace*{2cm}\=\hspace*{2cm}\= \kill
20       $\square$ 0.34 \> $\square$ 0.5 \> $\square$ 0.3 \\
21       \end{tabbing}
22       \begin{solution}[]
23           Poisson's ratio of steel is approximately 0.3.
24           This value is valid at moderate temperatures.
25       \end{solution}
26       \item\PTs{2}   Calculate $2+3 =\;$
27       \begin{solution}[]
28           Correct calculation: $2+3=5$.
29       \end{solution}
30       \end{parts}
31   \end{exercise*}
```

**Listing 6.31** Environment for an exercise with a parts environment for multiple parts questions

EXERCISE 1. Calculate the first-order derivatives of the following functions $f(x)$:

(a) $(2^{\text{pts}})$ $2 \times x^2$                                 (b) $(2^{\text{pts}})$ $\sin(x)$

(c) $(2^{\text{pts}})$ $\ln(x)$                                         (d) $(2^{\text{pts}})$ $\frac{1}{x}$

**Fig. 6.31** An exercise with a parts environment for multiple parts questions in tabular form, see Listing 6.32

Alternatively to Fig. 6.30, a multiple parts questions can be arranged in tabular form. The optional argument[9] of the **parts** environment allows to define the number of columns to be used. Columns are separated by the classical **&** and the end of the row is indicated by a \\, see Listing 6.32 and Figs. 6.31 and 6.32 for the corresponding outputs. However, it should be noted here that the optional parameter for the solution (i.e., the vertical space for the user to write the answer) is ignored for exercises with parts in tabular format.

```
1    ...
2    \usepackage[nosolutions]{exerquiz}
3    %\usepackage[solutionsafter]{exerquiz}
4    ...
5    \begin{document}
6    ...
7    \PTsHook{($\eqPTs^{\text{pts}}$)}
8    \begin{exercise*}
9    Calculate the first-order derivatives of the following functions $f(x)$:
10       \begin{parts}[2]
11       \item\PTs{2}       $2\times x^2$
12       \begin{solution}[]
13          Derivative $f'(x)=4\times x$.
14       \end{solution}
15       &
16       \item\PTs{2} $\sin(x)$
17       \begin{solution}[]
18          Derivative $f'(x)=\cos(x)$.
19       \end{solution}
20       \\
21       \item\PTs{2} $\ln(x)$
22       \begin{solution}[]
23          Derivative $f'(x)=\tfrac{1}{x}$.
24       \end{solution}
25       &
26       \item\PTs{2} $\tfrac{1}{x}$
27       \begin{solution}[]
28          Derivative $f'(x)=-\tfrac{1}{x^2}$.
29       \end{solution}\\
30       \end{parts}
31    \end{exercise*}
```

**Listing 6.32** Environment for an exercise with a parts environment for multiple parts questions in tabular form

Let us now focus a bit on redesigning the **exercise** environment. Listing 6.33 shows the example of defining a new environment, called **problem**, which is based on the **exercise\*** environment. The **\exlabel** command allows to replace the standard expression 'Exercise' and the **\exlabelformat** command allows the corresponding

---

[9] Without the optional parameter, a list environment is used.

EXERCISE 1. Calculate the first-order derivatives of the following functions $f(x)$:

(a) ($2^{\text{pts}}$) $2 \times x^2$                          (b) ($2^{\text{pts}}$) $\sin(x)$

    *Solution*: Derivative              *Solution*: Derivative

    $f'(x) = 4 \times x$.       □     $f'(x) = \cos(x)$.      □

(c) ($2^{\text{pts}}$) $\ln(x)$                          (d) ($2^{\text{pts}}$) $\frac{1}{x}$

    *Solution*: Derivative $f'(x) = \frac{1}{x}$. □   *Solution*: Derivative

                            $f'(x) = -\frac{1}{x^2}$.    □

**Fig. 6.32** An exercise with a parts environment for multiple parts questions in tabular form, see Listing 6.32

formatting, see Fig. 6.33 for the corresponding output. Table 6.3 collects some important formatting commands, see [38].

```
1   \usepackage[nosolutions]{exerquiz}
2   %\usepackage[solutionsafter]{exerquiz}
3   ...
4   \begin{document}
5   ...
6   \newenvironment{problem}{
7   \renewcommand\exlabel{Problem}
8   \renewcommand\exlabelformat{\textbf{\exlabel\ \theeqexno:}}
9   \begin{exercise*}}{\end{exercise*}}
10
11  \PTsHook{($\eqPTs^{\text{pts}}$)}
12  \begin{problem}
13   Calculate the first-order derivatives of the following functions $f(x)$:
14      \begin{parts}[2]
15      \item\PTs{2}     $2\times x^2$
16      \begin{solution}[]
17         Derivative $f'(x)=4\times x$.
18      \end{solution}
19      &
20      \item\PTs{2} $\sin(x)$
21      \begin{solution}[]
22         Derivative $f'(x)=\cos(x)$.
23      \end{solution}
24      \\
25      \item\PTs{2} $\ln(x)$
26      \begin{solution}[]
27         Derivative $f'(x)=\tfrac{1}{x}$.
28      \end{solution}
29      &
30      \item\PTs{2} $\tfrac{1}{x}$
31      \begin{solution}[]
32         Derivative $f'(x)=-\tfrac{1}{x^2}$.
33      \end{solution}\\
34      \end{parts}
35  \end{problem}
```

**Listing 6.33** Environment for an exercise with a redesigned layout

The variable **eqexno** is the counter for the exercise number. In case that no exercise number is required, the example of Listing 6.33 could be modified as shown in Listing 6.34.

**Table 6.3** Commands for changing the standard layout of the exercise environment

| Command | Comment |
|---------|---------|
| \exlabel | To replace the default word 'Exercise'. Default usage: \newcommand\exlabel{Exercise} |
| \exlabelformat | To change the formatting of the 'Exercise' label (e.g. small caps, boldface etc.). Default usage: \newcommand\exlabelformat{{\scshape\exlabel\theeqexno.}} |
| \exlabelsol | To change the name of the exercise label in the solutions section. Default usage: \newcommand\exlabelsol{\exlabel} |
| \exsllabelformat | To change the format of the solutions label. Default usage: \newcommand\exsllabelformat {\noexpand\textbf{\exlabelsol\\theeqexno.}} |
| \exrtnlabelformat | The label to click on to return from the solution. Default usage: \newcommand\exrtnlabelformat{\exlabelsol\\theeqexno} |
| \exsectitle | The section title of the solutions. Default usage: \newcommand\exsectitle{Solutions to \exlabel s} |
| \exsecrunhead | The running header for the solution section. Default usage: \newcommand\exsecrunhead{\exsectitle} |

**Problem 1:** Calculate the first-order derivatives of the following functions $f(x)$:

(a) $(2^{pts})$ $2 \times x^2$        (b) $(2^{pts})$ $\sin(x)$

(c) $(2^{pts})$ $\ln(x)$        (d) $(2^{pts})$ $\frac{1}{x}$

**Fig. 6.33** An exercise with a redesigned environment, see Listing 6.33

```
1   \usepackage[nosolutions]{exerquiz}
2   %\usepackage[solutionsafter]{exerquiz}
3   ...
4   \begin{document}
5   ...
6   \newenvironment{problem}{
7   \renewcommand\exlabel{Problem}
8   \renewcommand\exlabelformat{\textbf{\exlabel:}}
9   \begin{exercise*}}{\end{exercise*}}
10  ...
```

**Listing 6.34** Environment for an exercise without exercise number

In the common LATEX way (i.e., using the setcounter command), the counter eqexno can be set to another number, see Listing 6.35 and Fig. 6.34 for an example.

EXERCISE 6. Calculate the first-order derivatives of the following functions $f(x)$:

(a) $(2^{pts})$ $2 \times x^2$        (b) $(2^{pts})$ $\sin(x)$

(c) $(2^{pts})$ $\ln(x)$        (d) $(2^{pts})$ $\frac{1}{x}$

**Fig. 6.34** An exercise with a modified counter number, see Listing 6.35

```
1   \setcounter{eqexno}{5}
2
3   \PTsHook{($\eqPTs^{\text{pts}}$)}
4   \begin{exercise*}
5   Calculate the first-order derivatives of the following functions $f(x)$:
6       \begin{parts}[2]
7       \item\PTs{2}     $2\times x^2$
8       \begin{solution}[]
9           Derivative $f'(x)=4\times x$.
10      \end{solution}
11      &
12      \item\PTs{2} $\sin(x)$
13      \begin{solution}[]
14          Derivative $f'(x)=\cos(x)$.
15      \end{solution}
16      \\
17      \item\PTs{2} $\ln(x)$
18      \begin{solution}[]
19          Derivative $f'(x)=\tfrac{1}{x}$.
20      \end{solution}
21      &
22      \item\PTs{2} $\tfrac{1}{x}$
23      \begin{solution}[]
24          Derivative $f'(x)=-\tfrac{1}{x^2}$.
25      \end{solution}\\
26      \end{parts}
27  \end{exercise*}
```

**Listing 6.35** Environment for an exercise with a modified counter number

The following example (see Listing 6.36 and Fig. 6.35) shows the modification of the **exercise*** environment and the introduction of a new counter (here called 'questno') based on the \newcounter command.

Question 4. Calculate the first-order derivatives of the following functions $f(x)$:

(a) ($2^{\text{pts}}$) $2 \times x^2$

(b) ($2^{\text{pts}}$) $\sin(x)$

*Solution*: Derivative $f'(x) = 4 \times x$. □

*Solution*: Derivative $f'(x) = \cos(x)$. □

(c) ($2^{\text{pts}}$) $\ln(x)$

(d) ($2^{\text{pts}}$) $\frac{1}{x}$

*Solution*: Derivative $f'(x) = \frac{1}{x}$. □

*Solution*: Derivative $f'(x) = -\frac{1}{x^2}$. □

**Fig. 6.35** An exercise in a new environment with its own counter, see Listing 6.36

```
1   \usepackage[nosolutions]{exerquiz}
2   %\usepackage[solutionsafter]{exerquiz}
3
4   \newcounter{questno}
5   \newenvironment{question}{%
6   \renewcommand\exlabel{Question}
7   \renewcommand\exlabelformat
8   {\textbf{\exlabel\ \thequestno.}}
9   \SolutionsAfter
10  \begin{exercise*}[questno]}%
11  {\end{exercise*}
12  \SolutionsAtEnd}
13
14  \begin{document}
15  ...
16  \setcounter{questno}{3}
17
18  \PTsHook{($\eqPTs^{\text{pts}}$)}
19  \begin{question}
20  Calculate the first-order derivatives of the following functions $f(x)$:
21      \begin{parts}[2]
22      \item\PTs{2}      $2\times x^2$
23      \begin{solution}[]
24          Derivative $f'(x)=4\times x$.
25      \end{solution}
26      &
27      \item\PTs{2} $\sin(x)$
28      \begin{solution}[]
29          Derivative $f'(x)=\cos(x)$.
30      \end{solution}
31      \\
32      \item\PTs{2} $\ln(x)$
33      \begin{solution}[]
34          Derivative $f'(x)=\tfrac{1}{x}$.
35      \end{solution}
36      &
37      \item\PTs{2} $\tfrac{1}{x}$
38      \begin{solution}[]
39          Derivative $f'(x)=-\tfrac{1}{x^2}$.
40      \end{solution}\\
41      \end{parts}
42  \end{question}
```

**Listing 6.36**  Environment for an exercise in a new environment with its own counter

## 6.3  eCards

The package **eCards** allows to design electronic flash cards. It is based on the
AcroTEX eDucation Bundle and relies on the **web**, **exerquiz** and **insdljs** packages,
see [37] for details. The additional files to AcroTEX can be downloaded as a com-
pressed folder from the following link [11]. The files include the required **eCards**
package, the documentation, and the example file **ecardstst.tex**. The general structure for an eCard is shown in Listing 6.37. There is the main
**\begin{card}...\end{card}** environment. The first item within this structure is the
question itself. Then follows a **response** environment, which contains a **hint** and
the corresponding **answer**. The optional argument of the **card** environment is **hint**

or **nohint**. The argument **nohint** suppresses any included hint in the **response** environment. If no argument is provided, it is assumed that the questions contains a hint.

```
1   \begin{card}[hint|nohint]
2   ... question text ...
3       \begin{response}
4           \begin{hint}
5           ... hint text ...
6           \end{hint}
7           \begin{answer}
8           ... answer text ...
9           \end{answer}
10      \end{response}
11  \end{card}
```

**Listing 6.37** General environment for an eCard

The simplest way to design eCards is to modify the provided example file **ecard-stst.tex**. This example contains a 'navigation' panel on the right-hand side of the page (initiated by the option **rightpanel** of the **web** package), which contains in the upper part a logo, in the middle part some text, and in the lower part some navigation buttons. Listing 6.38 shows the example of an eCard with a multiple choice question. The layout of the example file was modified by changing the logo (see the command **\ecLogo{\includegraphics...** in line 10) and the text (see the command **\newcommand\aebLogo...** in line 12) in the navigation panel. The navigation buttons are as in the original example file. The multiple choice question itself is enclosed in a **\begin{multiChoice}...\end{multiChoice}** environment. The optional argument of this environment (here: 3) specifies the number of columns.

```
1   \documentclass{article}
2
3   \usepackage{amsmath}
4   \usepackage{graphicx}
5   % Use these three for on screen presentation.
6   \usepackage[execJS]{exerquiz}
7   \usepackage[dvipso,tight,nodirectory,rightpanel]{web}
8   \usepackage[memLogo]{eCards}
9   ...
10  \ecLogo{\includegraphics[scale=.14]{graphics/Esslingen_University_Logo.png}}
11
12  \newcommand\aebLogo{\parbox{2.15in}{\large \color{red}\textsl{eCards:
13  Different Questions}\\
14  \small\smash{\raisebox{3pt}{\color{blue}\textsl{Multiple choice, math- and
15  text fill-in}\hfill}}}}
16  ...
17  \begin{document}
18  ...
19  \begin{card}
20  What is the Poisson's ratio of steel?
21      \begin{multiChoice}{3}
22      \Ans0 0.34 &\Ans0 0.5 &  \Ans1 0.3
23      \end{multiChoice}
24      \begin{response}
25      \begin{hint}
26          The common range for engineering materials is: $0\le \nu \le 0.5$.
27      \end{hint}
28      \begin{answer}
29          Poisson's ratio of steel is approximately 0.3.
```

```
30          This value is valid at moderate temperatures.
31        \end{answer}
32        \end{response}
33    \end{card}
34    ...
35    \end{document}
```

**Listing 6.38**  Environment for an eCard with a multiple choice question

The eCard defined in Listing 6.38 produces three different pages, each in a different layout: the question, hint, and answer page as shown in Fig. 6.36 in their standard layouts. It should be noted here that it might be helpful in some cases to also include figures on the hint or answer page. This can be done with the common \includegraphics command.

In addition to multiple choice questions, one can easily define fill-in questions as already explained in Sect. 6.2.1 for math and text fill-in questions. Listing 6.38 and Fig. 6.37 show the example of a text fill-in question. The \RespBoxTxtPC command from Sect. 6.2.1 is now embedded in a \begin{fillIn}...\end{fillIn} environment.

```
1     \begin{card}
2     Which minister proposed in 1950 the European
3     Coal and Steel Community (ECSC)?\\[1ex]
4         \begin{fillIn}
5         \RespBoxTxtPC{3}[ECSC]{2}
6             [1]{Robert}
7             [2]{Schuman}
8         \end{fillIn}
9         \begin{response}
10        \begin{hint}\raggedright
11            It was the French foreign minister.
12        \end{hint}
13        \begin{answer}
14            Robert Schuman was the French foreign minister who proposed in 1950
15            the European Coal and Steel Community (ECSC).
16        \end{answer}
17        \end{response}
18    \end{card}
```

**Listing 6.39**  Environment for an eCard with a text fill-in question

Let us now focus on further modifications of the standard example file. The first item to be modified is the panel with a certain set of navigation buttons. Listing 6.40 and Fig. 6.38 show the example of an eCard with modified navigation buttons in the right-hand panel. The available navigations buttons and their default English declaration are summarized in Table 6.4.

```
1   ...
2   \ecLogo{\includegraphics[scale=.14]{graphics/Esslingen_University_Logo.png}}
3
4   \newcommand\aebLogo{\parbox{2.15in}{\large \color{red}\textsl{eCards:
5   Different Questions}\small\smash{\raisebox{3pt}{\color{blue}\textsl{Multiple
6   choice, math- and text fill-in}\hfill}}}}
7
8   \renewcommand\panelNaviGroup {
9       {%
10          \NextCard\\
11          \Hint\\
12          \Soln\\
13          \Home\\
14      }%
15  }
16
17  \begin{document}
18  ...
19
20  \begin{card}
21  What is the Poisson's ratio of steel?
22      \begin{multiChoice}{3}
23      \Ans0 0.34 &\Ans0 0.5 &  \Ans1 0.3
24      \end{multiChoice}
25      \begin{response}
26      \begin{hint}
27          The common range for engineering materials is: $0\le \nu \le 0.5$.
28      \end{hint}
29      \begin{answer}
30          Poisson's ratio of steel is approximately 0.3.
31          This value is valid at moderate temperatures.
32      \end{answer}
33      \end{response}
34  \end{card}
```

**Listing 6.40** Environment for an eCard with modified navigation buttons in the right-hand panel

The outer dimensions of the navigation buttons can be set by using the commands[10] \renewcommand\iconWidth{40pt} and \renewcommand\iconHeight{15pt}.

Let us come back now to the layout of the question, hint, and answer page, see Fig. 6.36. The wordings 'QUESTION', 'HINT', and 'ANSWER' can be changed by modifying the \headCard, \headHint, and \headAnswer commands. The background colors of the corresponding pages and the panel can be changed by modifying the \cardColor{vlightblue}, \hintColor{cornsilk}, \solnColor{webyellow}, and \panelBgColor{logoblue} commands.[11] The web.sty package includes some color definitions, which were particularly defined for web presentations, see Table 6.5 (Fig. 6.39).

---

[10] The shown numbers are the default settings in the example file.

[11] The arguments are the default settings of the example file. 'logoblue' is defined as \definecolor{logoblue}{rgb}{0,0,0.267}.

**Fig. 6.36** **a** Question,
**b** hint, and **c** answer page,
see Listing 6.38

**(a)**

### QUESTION

What is the Poisson's ratio of steel?

(a) 0.34          (b) 0.5          (c) 0.3

**(b)**

### HINT

The common range for engineering materials
is: $0 \leq \nu \leq 0.5$.

**(c)**

### ANSWER

Poisson's ratio of steel is approximately 0.3.
This value is valid at moderate temperatures.

QUESTION

Which minister proposed in 1950 the European Coal and Steel Community (ECSC)?

**Fig. 6.37**  An eCard with a text fill-in question, see Listing 6.39

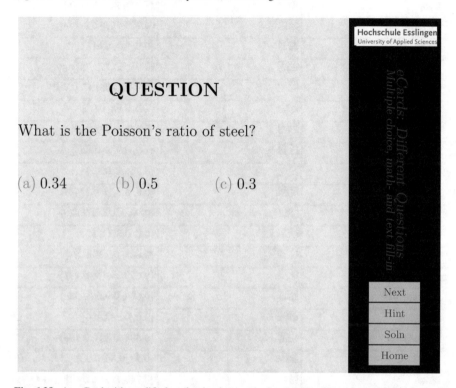

**Fig. 6.38**  An eCard with modified navigation buttons in the right-hand panel, see Listing 6.40

**Table 6.4**  Some available navigation buttons for eCards, see [11]

| Command | Default (English) declaration |
| --- | --- |
| \Soln | \ecSolnLabel{Soln} |
| \Hint | \ecHintLabel{Hint} |
| \NextCard | \ecNextLabel{Next} |
| \Home | \ecHomeLabel{Home} |
| \Close | \ecCloseLabel{Close} |
| \Begin | \ecBeginLabel{Begin} |

**Table 6.5** Predefined colors of the web style package

| Name | Display | rgb model |
|------|---------|-----------|
| webgreen | | {0.0, 0.5, 0.0} |
| webbrown | | {0.6, 0.0, 0.0} |
| webyellow | | {0.98, 0.92, 0.7} |
| webgray | | {0.753, 0.753, 0.753} |
| webgrey | | {0.753, 0.753, 0.753} |
| webblue | | {0.0, 0.0, 0.8} |
| wheat | | {0.96, 0.87, 0.70} |
| oldlace | | {0.992, 0.96187, 0.902} |
| snow | | {1.0, 0.98, 0.98} |
| ghostwhite | | {0.973, 0.973, 1.0} |
| cornsilk | | {1.0, 0.973, 0.863} |
| honeydew | | {0.941, 1.0, 0.941} |
| lavenderdark | | {0.8, 0.8, 0.9529411} |
| lavender | | {0.902, 0.902, 0.980} |
| lightblue | | {0.8, 0.8, 0.95} |
| lightgray | | {0.827, 0.827, 0.82} |
| lightsteelblue | | {0.690, 0.769, 0.871} |
| lightturquoise | | {0.686, 0.933, 0.933} |
| darkgreen | | {0.0, 0.392, 0.0} |
| yellowgreen | | {0.604, 0.804, 0.196} |
| vlightblue | | {0.88, 0.85, 0.95} |
| khaki | | {0.741, 0.718, 0.42} |

**Fig. 6.39** Modified text and color: **a** Question, **b** hint, and **c** answer page, see Listing 6.41

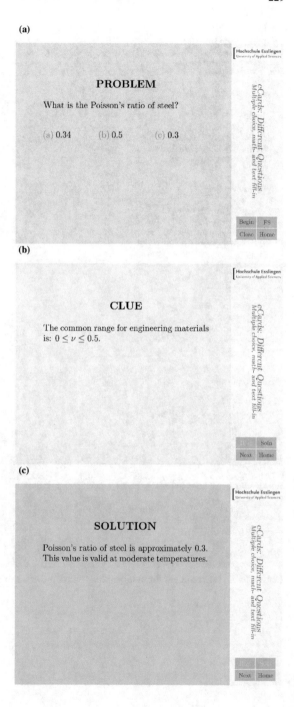

```
1   ...
2   \cardColor{wheat}
3   \hintColor{honeydew}
4   \solnColor{lightturquoise}
5   \panelBgColor{snow}
6
7   \renewcommand{\headCard}
8   {%
9       \vspace*{\stretch{.1}}%
10      \begin{center}%
11      \Large\textbf{PROBLEM}\par\vspace*{.25in}%
12      \begin{minipage}{.85\linewidth}%
13      \large\unskip\noindent\ignorespaces%
14  }
15
16  \renewcommand{\headHint}
17  {%
18      \vspace*{\stretch{.1}}%
19      \begin{center}%
20      \Large\textbf{CLUE}\par\vspace*{.25in}%
21      \begin{minipage}{.85\linewidth}%
22      \large\unskip\noindent\ignorespaces%
23  }
24
25  \renewcommand{\headAnswer}
26  {%
27      \vspace*{\stretch{.1}}%
28      \begin{center}%
29      \Large\textbf{SOLUTION}  \\\vspace*{.25in}%
30      \begin{minipage}{.85\linewidth}%
31      \large\unskip\noindent\ignorespaces%
32  }
33
34  \begin{document}
35  ...
36  \begin{card}
37  What is the Poisson's ratio of steel?
38      \begin{multiChoice}{3}
39      \Ans0 0.34 &\Ans0 0.5 &  \Ans1 0.3
40      \end{multiChoice}
41      \begin{response}
42      \begin{hint}
43          The common range for engineering materials is: $0\le \nu \le 0.5$.
44      \end{hint}
45      \begin{answer}
46          Poisson's ratio of steel is approximately 0.3.
47          This value is valid at moderate temperatures.
48      \end{answer}
49      \end{response}
50  \end{card}
51  ...
```

**Listing 6.41** Environment for modification of text and color: question, hint, and answer page

# Appendix A
# Color Definition in TikZ

Some of the common color models known in literature are mentioned in Table A.1.

In addition to the predefined colors in Table A.2, it is possible to mix colors. For example, the command red!20!white contains 20% of red and 80% of white and results in ▭. An example for the mixture of three colors may look like red!20!white!60!blue (▬) and mixes 20% red with 60% white and 20% blue.

The following example explains the use of the rgb color model with 95% red, 44% green and 13% blue.

```
\begin{tikzpicture}
\definecolor{color1}{rgb}{0.95,0.44,0.13}
\filldraw[fill=color1, draw=black] (0,0) rectangle (0.5,0.25);
\end{tikzpicture}
```

The use of the cmyk color model with 0% cyan, 70% magenta, 100% yellow and 0% black is shown in the following example.

```
\begin{tikzpicture}
\definecolor{color1}{cmyk}{0.0,0.7,1.0,0.0}
\filldraw[fill=color1, draw=black] (0,0) rectangle (0.5,0.25);
\end{tikzpicture}
```

■ ■

**Table A.1** Some common color models. Adapted from [19]

| Name | Base colors | Parameter range | Comment |
|---|---|---|---|
| rgb | red, green, blue | $[x, y, z]$ | From 0 to 1, floating point representation |
| RGB | Red, Green, Blue | $[x, y, z]$ | From 0 to 255, integer numbers |
| cmy | cyan, magenta, yellow | $[x, y, z]$ | From 0 to 1, floating point representation |
| cmyk | cyan, magenta, yellow, black | $[w, x, y, z]$ | From 0 to 1, floating point representation |

**Table A.2** Predefined colors in TikZ

| Name | Display | rgb model | cmyk system |
|---|---|---|---|
| red | | {1.0, 0.0, 0.0} | {0.0, 1.0, 1.0, 0.0} |
| green | | {0.0, 1.0, 0.0} | {1.0, 0.0, 1.0, 0.0} |
| blue | | {0.0, 0.0, 1.0} | {1.0, 1.0, 0.0, 0.0} |
| cyan | | {0.0, 1.0, 1.0} | {1.0, 0.0, 0.0, 0.0} |
| magenta | | {1.0, 0.0, 1.0} | {0.0, 1.0, 0.0, 0.0} |
| yellow | | {1.0, 1.0, 0.0} | {0.0, 0.0, 1.0, 0.0} |
| black | | {0.0, 0.0, 0.0} | {0.0, 0.0, 0.0, 1.0} |
| gray | | {0.5, 0.5, 0.5} | {0.0, 0.0, 0.0, 0.5} |
| darkgray | | {0.25, 0.25, 0.25} | {0.0, 0.0, 0.0, 0.75} |
| lightgray | | {0.75, 0.75, 0.75} | {0.0, 0.0, 0.0, 0.25} |
| brown | | {0.75, 0.5, 0.25} | {0.0, 0.25, 0.5, 0.25} |
| lime | | {0.75, 1.0, 0.0} | {0.25, 0.0, 1.0, 0.0} |
| olive | | {0.5, 0.5, 0.0} | {0.0, 0.0, 1.0, 0.5} |
| orange | | {1.0, 0.5, 0.0} | {0.0, 0.5, 1.0, 0.0} |
| pink | | {1.0, 0.75, 0.75} | {0.0, 0.25, 0.25, 0.0} |
| purple | | {0.75, 0.0, 0.25} | {0.0, 0.75, 0.5, 0.25} |
| teal | | {0.0, 0.5, 0.5} | {0.5, 0.0, 0.0, 0.5} |
| violet | | {0.5, 0.0, 0.5} | {0.0, 0.5, 0.0, 0.5} |
| white | | {1.0, 1.0, 1.0} | {0.0, 0.0, 0.0, 0.0} |

# Appendix B
# Units of Measure

The different units of measure in LATEX are summarized in Table B.1.

**Table B.1** The common units of measure in LATEX

| Unit | Name | Metric conversion |
|---|---|---|
| Font-independent | | |
| pt | point | 1 pt = 0.351459804 mm |
| pc | pica | 1 pc = 12 pt = 4.21751764 mm |
| bp | big point | 1 bp = 0.352777778 mm |
| dd | didot point | 1 dd = 0.376065028 mm |
| nd | new didot | 0.375 mm |
| cc | cicero | 4.51278034 mm |
| nc | new cicero | 4.5 mm |
| sp | scaled point | 1 sp = 5.36285101 $\times 10^{-6}$ mm |
| in | inch | 1 in = 25.4 mm |
| cm | centimeter | 1 cm = 10 mm |
| mm | millimeter | 1 mm |
| Font-dependent | | |
| ex | | a vertical measure usually about the height of the letter 'x' in the current font |
| em | | a horizontal measure usually about the width of the letter 'M' in the current font |
| mu | | math unit (18 mu = 1 em) for positioning in math mode |

M. Öchsner and A. Öchsner, *Advanced LaTeX in Academia*, https://doi.org/10.1007/978-3-030-88956-2

# References

1. Adriaens H (2017) The powerdot class. http://ctan.mines-albi.fr/macros/latex/contrib/powerdot/doc/powerdot.pdf. Cited 16 July 2020
2. Amberg B, Kainhofer R (2011) The baposter latex poster style. http://www.brian-amberg.de/uni/poster/baposter/baposter_guidepdf. Cited 20 August 2020
3. Beamerposter example 00 (2015). https://github.com/deselaers/latex-beamerposter/tree/master/examples/00. Cited 13 August 2020
4. Beamerposter example 01 (2015). https://github.com/deselaers/latex-beamerposter/tree/master/examples/01. Cited 13 August 2020
5. Beamerposter example 02 (2015). `https://github.com/deselaers/latex-beam-erposter/tree/master/examples/02`. Cited 19 August 2020
6. Beamer – A LaTeX class for producing presentations and slides (2020) https://ctan.org/pkg/beamer?lang=en. Cited 16 July 2020
7. Beamer theme gallery (2020). https://deic-web.uab.cat/~iblanes/beamer_gallery/index.html. Cited 15 July 2020
8. Beamer theme matrix (2020). https://hartwork.org/beamer-theme-matrix/. Cited 15 July 2020
9. Carlisle DP (2017) Packages in the 'graphics' bundle. https://www.ctan.org/pkg/graphicx. Cited 22 July 2020
10. Dreuw P, Deselaers T (2008) The beamerposter package. https://ctan.org/pkg/beamerposter. Cited 11 August 2020
11. ecards.zip (2016). https://ctan.org/pkg/ecards. Cited 3 March 2020
12. Feuersänger C (2017) Manual for package PGFPLOTS. https://ctan.org/pkg/pgfplots?lang=de. Cited 3 December 2017
13. Feuersänger C (2017) Manual for package PGFPLOTSTABLE. http://pgfplots.sourceforge.net/pgfplotstable.pdf. Cited 16 November 2017
14. Gnuplot homepage. http://www.gnuplot.info. Cited 14 December 2017
15. Goualard F, Neergaard PM (2003) Making slides in LaTeX with prosper. https://www.ctan.org/pkg/prosper. Cited 16 July 2020
16. Grahn A (2012) The movie15 package. https://www.ctan.org/pkg/movie15. Cited 16 July 2020
17. Grahn A (2020) The media9 package, v1.12. https://www.ctan.org/pkg/media9. Cited 16 July 2020
18. Hirschhorn P (2017) Using the exam document class. http://www-math.mit.edu/~psh/exam/examdoc.pdf. Cited 11 October 2019

© The Editor(s) (if applicable) and The Author(s), under exclusive license
to Springer Nature Switzerland AG 2021
M. Öchsner and A. Öchsner, *Advanced LaTeX in Academia*,
https://doi.org/10.1007/978-3-030-88956-2

19. Kern U (2016) Extending LaTeX's color facilities: the xcolor package. http://texdoc.net/
    texmf-dist/doc/latex/xcolor/xcolor.pdf. Cited 2 February 2020
20. Kettl G, Weiser M (2004) a0poster. http://mirror.physik-pool.tu-berlin.de/pub/CTAN/macros/
    latex/contrib/a0poster/a0_eng.pdf. Cited 11 August 2020
21. Kohm M (2020) The guide KOMA-script. https://www.ctan.org/pkg/scrartcl.
    Cited 16 July 2020
22. de Luna TM (2008) Writing posters in LaTeX. https://www.tug.org/pracjourn/2008-3/morales/
    morales.pdf. Cited 12 August 2020
23. Niederberger C (2019) The ExSheets bundle. http://ctan.math.washington.edu/tex-archive/
    macros/latex/contrib/exsheets/exsheets_en.pdf. Cited 13 October 2019
24. Öchsner A (2013) Introduction to scientific publishing: backgrounds, concepts. strategies.
    Springer, Heidelberg
25. Öchsner A (2014) Elasto-plasticity of frame structure elements: modeling and simulation of
    rods and beams. Springer, Berlin
26. Öchsner M, Öchsner A (2015) Das Textverarbeitungssystem LaTeX: Eine praktische Ein-
    führung in die Erstellung wissenschaftlicher Dokumente. Springer Vieweg, Wiesbaden
27. Öchsner A (2020) Computational statics and dynamics: an introduction based on the finite
    element method. Springer, Singapore
28. Pégourié-Gonnard M (2008) The xargs package. https://ctan.org/pkg/xargs?lang=en. Cited 27
    January 2018
29. Pichaureau P (2014) exercise.sty : a package to typeset exercises. http://ctan.space-pro.be/tex-
    archive/macros/latex/contrib/exercise/exercise.pdf. Cited 13 October 2019
30. Richter P, Botoeva E, Barnard R, Surmann D (2014) The tikzposter class. https://ctan.mirror.
    norbert-ruehl.de/graphics/pgf/contrib/tikzposter/tikzposter.pdf. Cited 11 August 2020
31. Robson AP (2015) The flowchart package: flowchart shapes for TikZ. https://ctan.org/pkg/
    flowchart?lang=de. Cited 22 January 2018
32. Shang HL (2012) Writing posters with beamerposter package in LaTeX. https://tug.org/
    pracjourn/2012-1/shang/shang.pdf. Cited 12 August 2020
33. ShareLaTeX Blog (2018). https://www.sharelatex.com/blog/2013/08/29/tikz-series-pt3.html.
    Cited 22 January 2018
34. Story DP (2011) The eqexam Package part of the AcroTEX eDucation Bundle. http://www.
    math.uakron.edu/~dpstory/eqexam/eqexamman.pdf. Cited 13 October 2019
35. Story DP (2016) How to install JavaScript support files. https://ctan.org/tex-archive/macros/
    latex/contrib/acrotex/doc?lang=de. Cited 24 February 2020
36. Story DP (2016) test_install.pdf. https://ctan.org/tex-archive/macros/latex/contrib/acrotex/
    examples. Cited 24 February 2020
37. Story DP (2016) eCards: electronic flash cards. https://ctan.math.illinois.edu/macros/latex/
    contrib/ecards/doc/ecardsman.pdf. Cited 3 March 2020
38. Story DP (2018) The AcroTEX eDucation Bundle (AeB). https://ctan.org/pkg/acrotex?
    lang=en. Cited 30 January 2018
39. Tantau T (2015) The TikZ and PGF packages: manual for version 3.0.1a. http://sourceforge.
    net/projects/pgf. Cited 19 December 2017
40. Tantau T, Joseph Wright J, Miletić V (2020) The beamer class: user guide for version 3.58.
    http://tug.ctan.org/macros/latex/contrib/beamer/doc/beameruserguide.pdf. Cited 20 July 2020
41. Tantau T (2020) Beamer examples: a-lecture. https://ctan.org/tex-archive/macros/latex/
    contrib/beamer/doc/examples/a-lecture. Cited 20 July 2020
42. Tellechea C (2018) spreadtab v0.4d user's manual. https://ctan.org/pkg/spreadtab?lang=de.
    Cited 3 January 2018
43. Weisstein EW (2017) Cube root. From MathWorld–a wolfram web resource. http://mathworld.
    wolfram.com/CubeRoot.html. Cited 7 November 2017

# Index

Printed in the United States
by Baker & Taylor Publisher Services